EXAMINING
THE BIG QUESTIONS OF
TIME

Presented by Laura Helmuth

THE
GREAT
COURSES®

SCIENTIFIC
AMERICAN.

4840 Westfields Boulevard | Suite 500 | Chantilly, Virginia | 20151-2299
[PHONE] 1.800.832.2412 | [FAX] 703.378.3819 | [WEB] www.thegreatcourses.com

LEADERSHIP

PAUL SUIJK	President & CEO
BRUCE G. WILLIS	Chief Financial Officer
CALE PRITCHETT	Chief Marketing Officer
JOSEPH PECKL	SVP, Marketing
JASON SMIGEL	VP, Product Development
MARK LEONARD	VP, Technology Services
DEBRA STORMS	VP, General Counsel
KONSTANTINE GELFOND	VP, Customer Engagement
KEVIN MANZEL	Sr. Director, Content Development
ANDREAS BURGSTALLER	Sr. Director, Brand Marketing & Innovation
KEVIN BARNHILL	Director of Creative
GAIL GLEESON	Director, Business Operations & Planning
LISA SIMPSON	Director, Digital Marketing

PRODUCTION TEAM

ALISHA REAY	Producer
BRANDON HOPKINS	Content Developer
BETSY SILBER	Associate Producer
TRISA BARNHILL	Graphic Artists
JAMES NIDEL	
DANIEL RODRIGUEZ	
BRIAN SCHUMACHER	
OWEN YOUNG	Managing Editor
AMY NESTOR	Preditor
ANDREW VOLPE	Editor
CHARLES GRAHAM	Assistant Editor
ROBERTO DE MORAES	Directors
ALEXIS DOTY	
CHRIS HOOTH	Audio Engineers
ED SALTZMAN	
RICK FLOWE	Camera Operator
PAUL SHEEHAN	Production Assistant

PUBLICATIONS TEAM

FARHAD HOSSAIN	Publications Manager
MARTIN STEGER	Sr. Copywriter
KATHRYN DAGLEY	Graphic Designer
JESSICA MULLINS	Proofreader
ERIKA ROBERTS	Publications Assistant
WILLIAM DOMANSKI	Transcript Editor

Copyright © The Teaching Company, 2021

Printed in the United States of America

This book is in copyright. All rights reserved. Without limiting the rights under copyright reserved above, no part of this publication may be reproduced, stored in or introduced into a retrieval system, or transmitted, in any form, or by any means (electronic, mechanical, photocopying, recording, or otherwise), without the prior written permission of The Teaching Company.

Laura Helmuth
Scientific American

Laura Helmuth is the editor in chief of *Scientific American*. She earned a PhD in Cognitive Neuroscience from the University of California, Berkeley, and a graduate certificate in Science Communication from the University of California, Santa Cruz.

Laura has served as the health, science, and environment editor for *The Washington Post*; director of digital news for National Geographic Partners; science and health editor for *Slate*; science editor for *Smithsonian Magazine*; and editor for *Science*. She is currently on the advisory boards for Spectrum, an autism news website, and SciLine, an organization that helps journalists find the best scientific sources. She also serves on the board of directors for *High Country News*, a magazine that covers the American West. She is a member of the National Academies of Sciences, Engineering, and Medicine's Standing Committee on Advancing Science Communication and was previously president of the National Association of Science Writers.

TABLE OF CONTENTS

INTRODUCTION

Presenter Biography i
About Our Partner iii
Course Scope 1

GUIDES

1	A Matter of Time	2
2	The Myth of the Beginning of Time	22
3	That Mysterious Flow	49
4	Is Time an Illusion?	70
5	Time Travel and the Twin Paradox	98
6	A Chronicle of Timekeeping	120
7	Atomic Clocks	152
8	Times of Our Lives	175
9	Remembering When	202
10	Inconstant Constants	221
11	Atoms of Space and Time	242
12	Could Time End?	267

SUPPLEMENTARY MATERIAL

Image Credits 287

ABOUT OUR PARTNER

Scientific American covers the advances in research and discovery that are changing our understanding of the world and shaping our lives. Founded 1845, it is the oldest continuously published magazine in the United States and now reaches more than 10 million people around the world each month through its website, print and digital editions, newsletters, and app. Authoritative and engaging features, news, opinion, and multimedia stories from journalists and expert authors—including more than 200 Nobel Prize winners—provide need-to-know coverage, insights, and illumination of the most important developments at the intersection of science and society. *Scientific American* is published by Springer Nature. As a research publisher, Springer Nature is home to other trusted brands, including Springer, Nature Research, BMC, and Palgrave Macmillan.

Based on several articles from "A Matter of Time"—a special edition of *Scientific American*—this course provides an overview of what the world's foremost thinkers and researchers have surmised about time as well as perplexing questions that remain. For more information on specific topics within this course, refer to the course scope on page 1.

SCIENTIFIC AMERICAN

EXAMINING
THE BIG QUESTIONS OF
TIME

This course's content was adapted from the special issue of *Scientific American* titled "A Matter of Time." The course draws on many of the issue's articles, and its first lesson provides a general overview of time, its units, and how we interpret it. Next, lessons cover topics such as the beginning of time (lesson 2), how time flows (lesson 3), and whether time is an illusion (lesson 4).

Then, the course turns to time travel and the practical difficulties and paradoxes presented by it (lesson 5). After that, the course examines timekeeping throughout past centuries (lesson 6) as well as modern efforts at ultraprecise methods of tracking time (lesson 7).

Lesson 8 focuses on how humans experience the passage of time—and how it affects us. Subsequent topics include how the mind organizes and interprets time (lesson 9) and the potentially inconstant nature of supposedly constant quantities (lesson 10).

The penultimate lesson covers quantum loop gravity, which presents the idea that discrete pieces form time and space. The course concludes with a lesson based on a haunting question: Could time end?

1
A MATTER OF TIME

Time defines us; it frames our experiences. We can't live, or understand our lives, without it. Intuitively, it seems to shape reality as an elemental aspect of our universe along with space. Yet the more we investigate it, the more elusive it becomes. This lesson looks at different interpretations of time, especially how ours have changed over the years. First, though, it discusses some common units of time.

WINDOWS OF TIME

Time is a powerful scientific tool. It's the way we measure duration, and as such it's a variable in almost every experiment scientists perform. As the following examples show, the units of time we commonly use range from the infinitesimally brief to the interminably long.

ATTOSECOND: A BILLIONTH OF A BILLIONTH OF A SECOND. The most fleeting events that scientists can clock are measured in attoseconds. Using high-speed lasers, researchers have created light pulses lasting just 53 attoseconds. Although the interval seems unimaginably brief, it's an aeon compared with the Planck time—about 10^{-43} second—which is believed to be the shortest possible duration.

FEMTOSECOND: A MILLIONTH OF A BILLIONTH OF A SECOND. An atom in a molecule typically completes a single vibration in 10 to 100 femtoseconds, with each being a million of a billionth of a second.

PICOSECOND: A THOUSANDTH OF A BILLIONTH OF A SECOND. The bottom quark, a rare subatomic particle created in high-energy accelerators, lasts for one picosecond before decaying. The average lifetime of a hydrogen bond between water molecules at room temperature is three picoseconds.

Lesson 1 | A Matter of Time

NANOSECOND: A BILLIONTH OF A SECOND. In one nanosecond, a beam of light shining through a vacuum will travel only 30 centimeters (just under a foot).

MICROSECOND: A MILLIONTH OF A SECOND. In a microsecond, that beam of light will have traveled 300 meters, about the length of three football fields.

MILLISECOND: A THOUSANDTH OF A SECOND. A normal camera flash lasts about a millisecond.

TENTH OF A SECOND. A blink of an eye is about a tenth of a second. The human ear needs this much time to discriminate an echo from the original sound.

SECOND. A healthy person's heartbeat lasts about one second. Traditionally, the second was the 60th part of the 24th part of a day, but science has given it a more precise definition: it's the duration of 9,192,631,770 cycles of one type of radiation produced by a cesium 133 atom.

MINUTE. In a minute, a shrew's heart beats 1,000 times. The average person can speak about 150 words or read about 250 words. When Mars is closest to Earth, sunlight reflected off Mars's surface reaches us in about four minutes.

HOUR. Reproducing cells generally take about an hour to divide in two. The Old Faithful geyser in Yellow Stone National Park erupts about every hour and a half. Light from Pluto arrives at Earth in five hours and 20 minutes.

DAY. For humans, the day is perhaps the most natural unit of time. Currently clocked at 23 hours, 56 minutes and 4.1 seconds, our planet's rotation is constantly slowing because of gravitational drag from the Moon and other influences. The human heart beats about 100,000 times in a day.

YEAR. In a year, Earth makes one circuit around the Sun and spins on its axis 365.26 times. It takes 4.3 years for light from Proxima Centauri, the closest star to our Sun, to reach Earth.

CENTURY. The Moon recedes from Earth by 3.8 meters each century. A baby born today has about a one in three chance of living to 100, and giant tortoises can live as long as 177 years.

A MILLION YEARS. After traveling for a million years, a spaceship moving at the speed of light would not yet be halfway to the Andromeda galaxy (which is 2.3 million light-years away). Because of the movement of Earth's tectonic plates, Los Angeles will creep about 40 kilometers north-northwest of its present location in a million years.

A BILLION YEARS. It took approximately a billion years for the newly formed Earth to cool, develop oceans, give birth to single-celled life, and exchange its carbon dioxide–rich early atmosphere for an oxygen-rich one.

CLOCKING CULTURES

- Social scientists have recorded wide differences in the pace of life in various countries and in how societies view time—whether as an arrow piercing the future or as a revolving wheel in which past, present, and future cycle endlessly. Some cultures even conflate time and space: The Australian Aborigines' concept of the Dreamtime encompasses not only a creation myth but a method of finding their way around the countryside.

- The study of time and society can be divided into the pragmatic and the cosmological. On the practical side, in the 1950s anthropologist Edward T. Hall Jr. wrote that the rules of social time constitute a "silent language" for a given culture. The rules might not always be made explicit, he stated, but "are either familiar and comfortable or unfamiliar and wrong."

- In 1955, he described in *Scientific American* how differing perceptions of time can lead to misunderstandings between people from separate cultures. He uses the example of a visitor to a foreign country keeping an ambassador waiting "for more than half an hour" and then "mutters an apology." This, per Hall, "is not necessarily an insult." Differences in time units between the two cultures may make the visitor not feel so late.

- Most cultures around the world now have clocks and calendars, uniting the majority of the globe in the same general rhythm of time. But that doesn't mean we all march to the same beat. Some people feel rushed by the pace of modern life, while in other societies, people feel little pressure to "manage" their time.

- Kevin K. Birth, an anthropologist, has examined time perceptions in Trinidad. Birth's 1999 book, *Any Time Is Trinidad Time: Social Meanings and Temporal Consciousness*, refers to a commonly used phrase to excuse lateness. In that country, Birth observes, "if you have a meeting at 6:00 at night, people show up at 6:45 or 7:00 and say, 'Any time is Trinidad time.'"

- When it comes to business, however, that loose approach to timeliness works only for the people with power. A boss can show up late and toss off "any time is Trinidad time," but underlings are expected to be more punctual.

- However, the nebulous nature of time can make it difficult for anthropologists and social psychologists to study. How people deal with time day-to-day often has nothing to do with how they conceive of time as an abstract entity.

▲ Some cultures don't draw neat distinctions among the past, present, and future. Aboriginal people in Australia, for instance, believe that their ancestors crawled out of the earth during the Dreamtime. The ancestors "sang" the world into existence as they moved about naming each feature and living thing. Even today an entity doesn't exist unless an Aborigine "sings" it.

REAL TIME

▲ More than 200 years ago, Benjamin Franklin equated passing minutes and hours with shillings and pounds, which has been shortened to the phrase "time is money." Time has become to the 21st century what fossil fuels and precious metals were to previous epochs. Constantly measured and priced, this vital raw material continues to spur the growth of economies built on a foundation of terabytes per second.

▲ Our commodification of time results from a radical alteration in how we view the passage of events. Our fundamental human drives haven't changed from the Paleolithic era: we have the same impulses to eat, procreate, fight, or flee. But in the long transition from then to now, our subjective experience of time has been completely transformed.

▲ By one definition, time is a continuum in which one event follows another from the past through to the future. But today the number of occurrences packed inside a given interval, be it a year or a nanosecond, increases relentlessly. The technological age has become a game in which more is always better.

- In his 2000 book *Faster: The Acceleration of Just About Everything*, James Gleick notes that before Federal Express shipping became commonplace in the 1980s, the exchange of business documents didn't usually require a package to be delivered "absolutely positively overnight."

- At first, FedEx gave its customers an edge. Soon, though, the whole world expected goods to arrive the next morning. "When everyone adopted overnight mail, equality was restored," Gleick writes, "and only the universally faster pace remained."

- The advent of the internet eliminated the burden of having to wait until the next day for a delivery. In internet time, everything happens everywhere at once: Connected computer users can witness an update to a web page at an identical moment in New York City or Dakar. Time has, in essence, triumphed over space.

> **ABOUT THIS LESSON**
>
> This lesson is largely based on the introduction to the *Scientific American* special issue "A Matter of Time" and an overview of time units created by David Labrador for the same edition. Additional source articles within the issue include "Clocking Cultures" by *Scientific American*'s editors and "Real Time" by Gary Stix.

Lesson 1 Transcript
A MATTER OF TIME

What is time? Well, that's a much harder question to answer than you might think. Time defines us; it frames our experiences. We can't live, or understand our lives, without it. Intuitively, it seems to shape reality as an elemental aspect of our universe along with space. Yet the more we investigate time, the more elusive it becomes—to the point that some scientists and philosophers believe it may be nothing more than an illusion.

I'm Laura Helmuth, editor in chief of *Scientific American*, and I'm about to take you on a tour of some of the best articles on time the magazine has published in the past two decades. These pieces were originally collected in a special issue called *A Matter of Time*; they examine various aspects of temporality through the lens of physics and psychology, philosophy and cosmology, neuroscience, clockwork mechanics, and more.

Over the next six hours, you'll discover just what a mind-blowing matter time truly is. But let's begin with a more straightforward definition: Time is a powerful scientific tool. It's the way we measure duration and as such it's a variable in almost every experiment scientists ever perform. And the units of time range from the infinitesimally brief to the interminably long.

The most fleeting events that scientists can clock are measured in attoseconds. Using high-speed lasers, researchers have created light pulses lasting just 53 attoseconds. Although the interval seems unimaginably brief, it's an aeon compared with the Planck time, which is about 10 to the negative-43 second—which is believed to be the shortest possible duration.

An atom in a molecule typically completes a single vibration in 10 to 100 femtoseconds. Even fast chemical reactions generally take hundreds of femtoseconds to complete. The interaction of light with the pigments in the retina—the process that allows vision—takes about 200 femtoseconds.

The bottom quark, a rare subatomic particle created in high-energy accelerators, lasts for one picosecond before decaying. The average lifetime of a hydrogen bond between water molecules at room temperature is three picoseconds.

In one nanosecond, a beam of light shining through a vacuum will travel only 30 centimeters (which is just under a foot). The microprocessor inside a personal computer will take less than a nanosecond to execute a single instruction, such as adding two numbers. The K meson, another rare subatomic particle, has a lifetime of 12 nanoseconds.

In a microsecond, a beam of light will have traveled about 300 meters, or about the length of three football fields, but a sound wave at sea level will have propagated only a third of a millimeter. The flash of a highspeed commercial stroboscope lasts about one microsecond. It takes about 24 microseconds for a stick of dynamite to explode after its fuse has burned down.

A normal camera flash lasts about a millisecond. A housefly flaps its wings once every three milliseconds; a honeybee, once every five milliseconds. The moon travels around Earth two milliseconds more slowly each year as its orbit gradually widens. In computer science, an interval of 10 milliseconds is known as a jiffy.

A "blink of an eye" is about one-tenth of a second. The human ear needs this much time to discriminate an echo from the original sound. *Voyager* 1, a spacecraft speeding out of the solar system, travels about two kilometers farther away from the sun during this time frame. A tuning fork pitched to A above middle C vibrates 44 times.

A healthy person's heartbeat lasts about one second. Earth travels 30 kilometers around the Sun, while the Sun zips 274 kilometers on its trek through the galaxy. A second isn't quite enough time for moonlight to reach Earth (that takes 1.3 seconds).

Traditionally, the second was the 60th part of the 24th part of a day, but science has given it a more precise definition: it's the duration of 9,192,631,770 cycles of one type of radiation produced by a cesium 133 atom.

In a minute, the brain of a newborn baby grows one to two milligrams. A shrew's fluttering heart beats 1,000 times. The average person can speak about 150 words or read about 250 words. When Mars is closest to Earth, sunlight reflected off the Red Planet's surface reaches us in about four minutes.

Reproducing cells generally take about an hour to divide in two. The Old Faithful geyser in Yellowstone National Park erupts about every hour and a half. Light from Pluto arrives at Earth in five hours and 20 minutes.

For humans, the day is perhaps the most natural unit of time. Currently clocked at 23 hours, 56 minutes and 4.1 seconds, our planet's rotation is constantly slowing because of gravitational drag from the Moon and other influences. The human heart beats about 100,000 times in a day and the lungs inhale about 11,000 liters of air. In the same amount of time, an infant blue whale adds another 91 kilograms (or about 200 pounds) to its bulk.

In a year, Earth makes one circuit around the sun and spins on its axis 365.26 times. North America moves about three centimeters away from Europe. It takes 4.3 years for light from Proxima Centauri, the closest star, to reach the Earth; and that is approximately the same amount of time that ocean-surface currents take to circumnavigate the globe.

The Moon recedes from Earth by another 3.8 meters each century. A baby born today has about a one in three chance of living to 100, and giant tortoises can live as long as 177 years.

After traveling for a million years, a spaceship moving at the speed of light would be not yet halfway to the Andromeda galaxy (which is 2.3 million light-years away). The most massive stars, blue supergiants that are millions of times brighter than the Sun, burn out in about this much time. Because of the movement of Earth's tectonic plates, Los Angeles will creep about 40 kilometers north-northwest of its present location in a million years.

It took approximately a billion years for the newly formed Earth to cool, develop oceans, and give birth to single-celled life and exchange its carbon dioxide–rich early atmosphere for an oxygen-rich one. Meanwhile, the Sun orbited four times around the center of the galaxy.

Because the universe is almost 14 billion years old, units of time beyond a billion years aren't used very often. But the universe will keep expanding indefinitely, until long after the last star dies (which will be about 100 trillion years from now) and the last black hole evaporates (which will be about 10 to the 100th power years from now). Our future stretches ahead much further than our past trails behind.

When we ask, "What is time?" the answer also varies from society to society. Show up an hour late in Brazil and no one bats an eyelash. But keep someone in Switzerland waiting for five or 10 minutes and you'll have some explaining to do. Time is elastic in many cultures but snaps taut in others. Indeed, the way members of a culture perceive and use time reflects their society's priorities and worldview.

Social scientists have recorded wide differences in the pace of life in various countries and in how societies view time—whether as an arrow piercing the future or as a revolving wheel in which past, present and future cycle endlessly. Some cultures even conflate time and space. In Australia, the Aboriginal people's concept of Dreamtime encompasses not only a creation myth but a method of finding their way around the countryside.

The study of time and society can be divided into the pragmatic and the cosmological. On the practical side, in the 1950s anthropologist Edward T. Hall wrote that the rules of social time constitute a "silent language" for a given culture. The rules might not always be made explicit, he stated, but "they always exist in the air. … They are either familiar and comfortable or unfamiliar and wrong."

In 1955, he described in *Scientific American* how differing perceptions of time can lead to misunderstandings between people from different cultures. "An ambassador who has been kept waiting for more than half an hour by a foreign visitor needs to understand that if his visitor 'just mutters an apology' this is not necessarily an insult," Hall wrote. "The time system in the foreign country may be composed of different basic units, so that the visitor is not as late as he may appear to be to us. You must know the time system of the country to know at what point apologies are really due. ... Different cultures simply place different values on the time units."

Most cultures around the world now have clocks and calendars, so uniting the majority of the globe in the same general rhythm of time. But that doesn't mean we all march to the same beat. Some people may feel so rushed by the pace of modern life that they're fighting back with "slow food," while in other societies, people feel little pressure to "manage" their time.

"One of the beauties of studying time is that it's a wonderful window on culture," says Robert Levine, a social psychologist at California State University, Fresno. "You get answers on what cultures value and believe in. You get a really good idea of what's important to people," he says.

Kevin K. Birth, an anthropologist at Queens College, has examined time perceptions in Trinidad. Birth's 1999 book, *Any Time Is Trinidad Time: Social Meanings and Temporal Consciousness*, he refers to a commonly used phrase to excuse lateness. In that country, Birth says, "If you have a meeting at 6:00 at night, people show up at 6:45 or 7:00 and say, 'Any time is Trinidad time.'" But when it comes to business, that loose approach to timeliness works only for the people with power. A boss can show up late and toss off "any time is Trinidad time," but underlings are expected to be more punctual. For them, the saying goes, "time is time." Birth adds that the tie between power and waiting time is true for many other cultures as well.

But the nebulous nature of time can make it a difficult thing for anthropologists and social psychologists to study. "You can't simply go into a society, walk up to some poor soul and say, 'Tell me about your notions of time,'" Birth says. "People don't really have an answer to that. You have to come up with other ways to find out."

How people deal with time day-to-day often has nothing to do with how they conceive of time as an abstract entity. "There's often a disjunction between how a culture views the mythology of time and how [people] think about time in their daily lives," Birth asserts. "We don't think of Stephen Hawking's theories as we go about our daily lives."

Some cultures don't draw neat distinctions among the past, present and future. Aboriginal people in Australia, for instance, believe that their ancestors crawled out of the earth during the Dreamtime. The ancestors "sang" the world into existence as they moved about naming each feature and living thing. Even today an entity doesn't exist unless an Aborigine person "sings" it.

Ziauddin Sardar, a British Muslim author and critic, has written about time and Islamic cultures, particularly the fundamentalist sect Wahhabism. Muslims "always carry the past with them," claims Sardar. "In Islam, time is a tapestry incorporating the past, present and future. The past is ever present." The followers of Wahhabism, which is widely practiced in Saudi Arabia, seek to re-create the idyllic days of the Prophet Muhammad's life. "The worldly future dimension has been suppressed" by them, Sardar says. "They have romanticized a particular vision of the past. All they are doing is trying to replicate that past."

But Sardar also asserts that the West has "colonized" time by spreading the expectation that life should become better as time passes. "If you colonize time, you also colonize the future," he says. "If you think of time as an arrow, of course you think of the future as progress, going in one direction. But different people may desire different futures."

Throughout this series, we'll explore topics drawn directly from the pages of *Scientific American* magazine. One such piece is called "Real Time." It is by senior editor Gary Stix, which provides a sneak peek of some of the themes we'll return to later.

More than 200 years ago, Benjamin Franklin equated passing minutes and hours with shillings and pounds, which has been shortened to "time is money." Time has become to the 21st century what fossil fuels and precious metals were to previous epochs. Constantly measured and priced, this vital raw material continues to spur the growth of economies built on a foundation of terabytes per second.

Our commodification of time results from a radical alteration in how we view the passage of events. Our fundamental human drives haven't changed from the Paleolithic era: we have the same impulses to eat, procreate, fight or flee that motivated Fred Flintstone. But in the long transition from Stone Age to Information Age, our subjective experience of time has been completely transformed.

By one definition, time is a continuum in which one event follows another from the past through to the future. But today the number of occurrences packed inside a given interval, be it a year or a nanosecond, increases relentlessly. The technological age has become a game in which more is always better.

In his book *Faster: The Acceleration of Just About Everything,* James Gleick notes that before Federal Express shipping became commonplace in the 1980s, the exchange of business documents didn't usually require a package to be delivered "absolutely positively overnight." At first, FedEx gave its customers an edge. But, soon though, the whole world expected goods to arrive the next morning. "When everyone adopted overnight mail, equality was restored," Gleick writes, "and only the universally faster pace remained."

The advent of the internet eliminated the burden of having to wait until the next day for the FedEx truck. In internet time, everything happens everywhere at once; connected computer users can witness an update to a web page at an identical moment in New York City or Dakar. Time has, in essence, triumphed over space.

Lesson 1 Transcript | A Matter of Time

The wired world erases time barriers and this achievement relies on an ever progressing ability to measure time more precisely. Over the eons, the capacity to gauge duration has correlated directly with increasing control over the environment that we inhabit. Keeping time is a practice that may go back more than 20,000 years, when hunters of the Ice Age notched holes in sticks or bones, possibly to track the days between phases of the Moon. A mere 5,000 years ago or so, the Babylonians and Egyptians devised calendars for planting and other time-sensitive activities.

Early chrono-technologists tracked natural cycles: the solar day, the lunar month and the solar year. The sundial could do little more than cast a shadow, when clouds or night didn't render it a useless decoration. But beginning in the 13th century, the mechanical clock initiated a revolution equivalent to the one engendered by the Gutenberg of the printing press. Time no longer "flowed," as it did literally in a water clock. Rather it was marked off by a mechanism that could track the beats of an oscillator. When refined, this device let time's passage be counted to fractions of a second.

The mechanical clock ultimately enabled the miniaturization of the timepiece. Once it was driven by a coiled spring and not a falling weight, it could be carried or worn like jewelry. The technology changed our perception of the way society was organized. It was an instrument that let one person coordinate activities with another. As Harvard University historian David S. Landes writes in his book *Revolution in Time*: "It is the mechanical clock that made possible, for better or worse, a civilization attentive to the passage of time, hence to productivity and performance."

Mechanical clocks persisted as the most accurate timekeepers for centuries. But the past 50 years have seen as much progress in the quest for precision as in the previous 700. It's not just the internet that brought about the conquest of time over space. Time is more accurately measured than any other physical entity, and as such, elapsed time is used to size up spatial dimensions. Today, standard makers gauge the length of the venerable meter by the distance that light travels in a vacuum in 1/299,792,458th of a second.

The atomic clocks used to make such measurements also play a role in judging location. In some of them, the resonant frequency of cesium atoms remains amazingly stable, becoming a pseudo pendulum capable of maintaining near nanosecond precision.

Global Positioning System satellites continuously broadcast their exact whereabouts as well as the time maintained by their onboard atomic clocks. A receiving device processes this information from at least four satellites into exact terrestrial coordinates, and these can be used by the pilot or a hiker, whether they're in Patagonia or Lapland. The requirements are exacting. A time error of only a millionth of a second from an individual satellite could send a signal to a GPS receiver that would be inaccurate by as much as a fifth of a mile.

Advances in precision timekeeping continue apace. In fact, clockmakers may soon outdo themselves. They may create an atomic clock so precise that it will be impossible to synchronize other timepieces to it. Researchers also continue to press ahead in slicing and dicing the second more finely. The need for speed has become a cornerstone of the Information Age. The world's speediest transistors can switch faster than a picosecond, a thousandth of a billionth of a second.

The modern era has also registered gains in assessing big intervals. Radiometric dating methods, measuring rods of "deep time," indicate how old the Earth really is. Just how far speed limits can be stretched remains to be tested. In conference sessions and popular books, physicists and theorists toy with ideas for the ultimate cosmic hot rod, a means of traveling forward or back in time.

Perplexity about the nature of time—it's a tripartite oddity that parses into past, present and future—precedes the industrial era by quite a few centuries. Augustine described the definitional dilemma more eloquently than anyone.

"What, then, is time?" he asked in his *Confessions*. "If no one asks me, I know; if I want to explain it to someone who does ask me, I do not know." He then went on to attempt to articulate why temporality is so hard to define: "How, then, can these two kinds of time, the past and the future, be, when the past no longer is and the future as yet does not be?"

Hard-boiled physicists, unburdened by theistic encumbrances, have also had difficulty grappling with this question. We remark that time "flies" as we hurtle toward our inevitable demise. What does that mean exactly? One could hypothesize a metric of current flow for time, a form of temporal amperage. But such a measure may simply not exist.

In fact, one of the hottest themes in theoretical physics is whether time itself is illusory. The confusion is such that physicists have gone as far as to recruit philosophers in their attempt to understand whether a time variable called t should be added to their equations.

The distinct feeling we have of being bookended between a past and a future—or perhaps being enmeshed in recurring natural rhythms—may be related to a basic biological reality. Our bodies are chock-full of living clocks—ones that govern how we connect a ball with a bat, when we feel sleepy and perhaps even when our time is up.

These real biorhythms have now begun to reveal themselves. Scientists are closing in on areas of the brain that produce the sensation of time flying when we're having fun. They're also beginning to understand the connections between different kinds of memory and how events are organized and recalled chronologically. Studies of neurological patients with various forms of amnesia, some of whom have lost the ability to judge accurately the passage of hours, months and even decades, are helping pinpoint which areas of the brain are involved in how we experience time.

Recalling where we fit in the order of things determines who we are. Ultimately, it doesn't matter whether time, in cosmological terms, retains an underlying physical truth. If it's a fantasy, it's one we cling to steadfastly. The reverence we hold for the fourth dimension, the complement of the three spatial ones, has much to do with a deep need to embrace meaningful temporal milestones that we can all share, like birthdays, holidays, anniversaries. And doing so seems to be the only way of imposing hierarchy and structure on a world that sometimes seems to lack any sense of permanence.

So what time is it? That simple question is probably asked more often today than ever. In the smartphone era, the answer's never more than a glance away, and so we can blissfully partition our days into ever smaller increments for ever more tightly scheduled tasks, confident that we'll always know it's, say, 11:55 pm.

However, modern scientific revelations about time make the question endlessly frustrating. If we seek a precise knowledge of the time, the elusive infinitesimal of *now* dissolves into a scattering flock of nanoseconds. Bound by the speed of light and the velocity of nerve impulses, our perceptions of the present sketch the world as it was an instant ago—for all our consciousness pretends otherwise, we can never catch up.

Even in principle, perfect synchronicity escapes us. Relativity dictates that, like a strange syrup, time flows slower on moving trains than in the train stations and faster in the mountains than it does in the valleys. The time for your wristwatch or digital screen isn't exactly the same as the time for your head. It's roughly 11:56 pm.

Our intuitions are deeply paradoxical. Time heals all wounds, but it's also the great destroyer. Time is relative but also relentless. There's time for every purpose under heaven, but there's never enough. Time flies, crawls and races. Seconds can be both split and stretched. Like the tide, time waits for no one, but in dramatic moments it also stands still. It's as personal as the pace of one's heartbeat but as public as the clock tower in the town square. We do our best to reconcile contradictions. It seems like 11:57 pm.

And of course, time is money. It's the partner of change, the antagonist of speed, the currency in which we pay attention. It's our most precious, irreplaceable commodity. Yet still we say we don't know where it goes, we sleep away a third of it, and none of us really can account for how much we have left. We can find 100 ways to save time, but the balance remaining nonetheless diminishes steadily. In fact, it's already 11:58 pm.

Time and memory shape our perceptions of our own identity. We may feel ourselves to be at history's mercy, yet we also see ourselves as free-willed agents of the future. But that conception is disturbingly at odds with the ideas of physicists and philosophers, for if time's a dimension like those of space, then yesterday, today and tomorrow are all equally concrete and determined. The future exists as much as the past does; it's just in a place we have yet to visit. Somewhere it's 11:59 pm.

"Time is the substance of which I am made," wrote Argentine author Jorge Luis Borges. "Time is a river that carries me away, but I am the river; it is a tiger that destroys me, but I am the tiger; it is a fire that consumes me, but I am the fire."

I hope you'll continue with me on this fascinating tour of what science has discovered about how time permeates and guides both our physical world and our inner selves. That knowledge enriches the imagination and provides practical advantages to anyone hoping to beat the clock or at least to stay in step with it.

It's now 12:00 am. Synchronize your watches.

2
THE **MYTH** OF THE **BEGINNING** OF TIME

Lesson 2 | The Myth of the Beginning of Time

Was the big bang really the beginning of time, or did the universe exist before then? That question is the focus of this lesson, and it is a question that seemed almost blasphemous only decades ago. Most cosmologists insisted that it made no sense. But developments in theoretical physics, especially the rise of string theory, have changed their perspective. The pre-bang universe has become a frontier of cosmology.

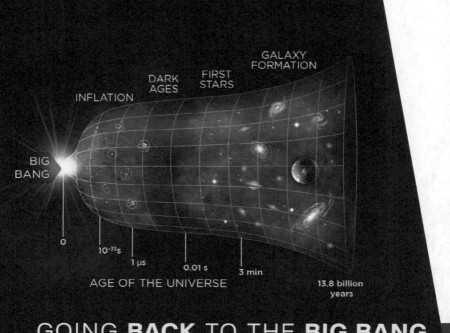

GOING BACK TO THE BIG BANG

◢ Einstein's general theory of relativity holds that space and time are malleable. On the largest scales, space is naturally dynamic, expanding or contracting over time, carrying matter like driftwood on the tide.

◢ In the 1920s, astronomers confirmed that our universe is currently expanding: Distant galaxies move apart from one another. One consequence, as physicists Stephen Hawking and Roger Penrose proved in the 1960s, is that time can't extend back indefinitely.

◢ As cosmic history plays backward in time, the galaxies all come together to a single infinitesimal point, known as a singularity. Each galaxy or its precursor is squeezed down to zero size. Quantities such as density, temperature, and spacetime curvature become infinite. The singularity is the ultimate cataclysm, beyond which our cosmic ancestry cannot extend.

Lesson 2 | The Myth of the Beginning of Time

- The unavoidable singularity poses serious problems for cosmologists. For the cosmos to look broadly the same everywhere, some kind of communication had to pass among distant regions of space, coordinating their properties. Yet the idea of such communication contradicts the standard understanding of cosmology.

- Consider what has happened over the 13.8 billion years since the big bang. The distance between galaxies has grown by a factor of about 1,000 because the universe is expanding, while the radius of the observable universe has grown by the much larger factor of about 100,000 because light outpaces the expansion.

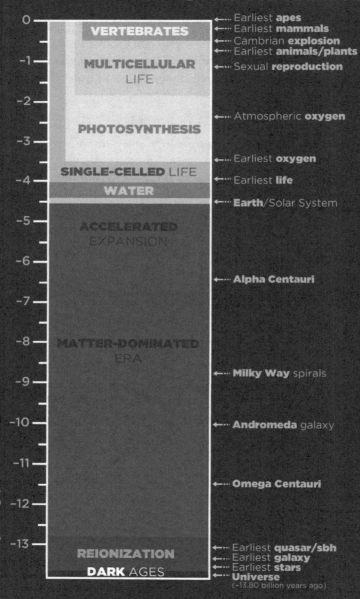

Lesson 2 | The Myth of the Beginning of Time

IDENTICAL REGIONS OF SPACE

- We see parts of the universe today that we couldn't have seen 13.8 billion years ago. Indeed, this is the first time in cosmic history that light from the most distant galaxies has reached the Milky Way. Nevertheless, the properties of the Milky Way are basically the same as those of distant galaxies.

- Here is an analogy: A person shows up to a party to find he's dressed in the exact same outfit as a dozen of his friends. This suggests that everyone coordinated their attire in advance. In cosmology, it's not clothes but tens of thousands of independent yet statistically identical patches of sky in the cosmic microwave background radiation, which is the afterglow from the big bang.

- One possible explanation is that all those regions of space were endowed at birth with identical properties. However, physicists have come up with two more natural ways out of the impasse: that the early universe was either much smaller or much older than in standard cosmology. Either (or both, acting together) would have made intercommunication possible.

- One explanation follows the first alternative. It postulates that the universe went through a period of accelerating expansion, known as inflation, early in its history. Before this phase, galaxies or their precursors were so close that they could easily coordinate their properties.

- During inflation, they fell out of contact because light couldn't keep pace with the frenetic expansion. After inflation ended, the expansion began to decelerate, so galaxies gradually came back into one another's view.

- Proposed in 1981, inflation has explained a wide variety of observations with precision. But a number of possible theoretical problems remain, beginning with the questions of what exactly the inflation was and what gave it such a huge initial potential energy.

- Another way to solve the puzzle follows the second alternative by getting rid of the singularity. If time didn't begin at the big bang, and if a long era preceded the onset of the present cosmic expansion, then matter could have had plenty of time to arrange itself smoothly. This has led some researchers to reexamine the reasoning that led them to infer a singularity.

- One of the assumptions—that relativity theory is always valid—is questionable. Close to the putative singularity, quantum effects must have been important. Standard relativity takes no account of such effects, so accepting the inevitability of the singularity amounts to trusting the theory beyond reason.

Lesson 2 | The Myth of the Beginning of Time

STRING THEORY

- To know what really happened, physicists need to include relativity in a quantum theory of gravity. Today, two approaches stand out. One is called loop quantum gravity. The second approach is string theory—a truly revolutionary modification of Einstein's theory. String theory is this lesson's focus, although proponents of loop quantum gravity reach many of the same conclusions.

- String theory grew out of a model that physicist Gabriele Veneziano wrote in 1968 to describe the world of nuclear particles (such as protons and neutrons) and their interactions. Despite much initial excitement, the model was abandoned several years later in favor of quantum chromodynamics, which describes nuclear particles in terms of more elementary constituents called quarks.

- Quarks are confined inside a proton or a neutron, as if they were tied together by elastic strings. In retrospect, the original string theory had captured those stringy aspects of the nuclear world. Only later was it revived as a candidate for combining general relativity and quantum theory.

- The basic idea is that elementary particles are not point-like but rather infinitely thin, one-dimensional objects—the strings. The large zoo of elementary particles, each with its own characteristic properties, reflects the many possible vibration patterns of a string.

QUANTUM STRING **MAGIC**

▲ How can such a "simple-minded theory," as Veneziano calls it, describe the complicated world of particles and their interactions? The answer can be found in what he dubbed "quantum string magic." Once the rules of quantum mechanics are applied to a vibrating string—just like a miniature violin string, except that the vibrations propagate along it at the speed of light—new properties appear. All have profound implications for particle physics and cosmology.

▲ First, quantum strings have a finite size. The irreducible quantum of length, denoted as l_s, plays a crucial role in string theory, putting a finite limit on quantities that otherwise could become either zero or infinite.

▲ Second, quantum strings may have angular momentum even if they lack mass. In classical physics, a massless object can have no angular momentum. But quantum fluctuations change the situation. A tiny string can acquire a certain amount of angular momentum without gaining any mass. Angular momentum clued physicists in to the quantum-gravitational implications of string theory.

▲ Third, quantum strings demand the existence of extra dimensions of space. The equations describing the quantum string vibrations become inconsistent unless spacetime is either highly curved (contrary to observations) or contains six additional spatial dimensions.

- Fourth, physical constants no longer have arbitrary, fixed values. They occur in string theory as fields that can adjust their values dynamically.

  - These fields may have taken different values in different cosmological epochs or in remote regions of space, and even today the physical "constants" may vary by a small amount. Observing any variation would provide an enormous boost to string theory.

  - One such field, called the dilaton, is the master key to string theory; it determines the overall strength of all interactions. The dilaton fascinates string theorists because its value can be reinterpreted as the size of an extra dimension of space, giving a grand total of 11 spacetime dimensions.

- Finally, quantum strings have introduced physicists to some striking new symmetries of nature known as dualities, which alter our intuition for what happens when objects get extremely small. Typically, a short string is lighter than a long one, but if we attempt to squeeze down its size below the fundamental length l_s, the string becomes heavier again.

T-DUALITY

- Another form of the symmetry, T-duality, holds that small and large extra dimensions are equivalent. This symmetry arises because strings can move in more complicated ways than point-like particles can. Consider a closed string (a loop) located on a cylindrically shaped space, whose circular cross-section represents one finite extra dimension. Besides vibrating, the string can either turn as a whole around the cylinder or wind around it.

- The energetic cost of these two states of the string depends on the size of the cylinder. The energy of winding is directly proportional to the cylinder radius. The energy associated with moving around the circle, on the other hand, is inversely proportional to the radius.

- If a large cylinder is substituted for a small one, the two states of motion can swap roles. Energies that had been produced by circular motion are instead produced by winding, and vice versa. An outside observer notices only the energy levels, not the origin of those levels. To that observer, the large and small radii are physically equivalent.

- Although T-duality is usually described in terms of cylindrical spaces, where one dimension (the circumference) is finite, a variant of it applies to our ordinary three dimensions, which appear to stretch on indefinitely. An infinite space's overall size can't change; it remains infinite. But it can still expand in the sense that bodies embedded within it, such as galaxies, move apart from one another.

- The crucial variable is not the size of the space as a whole but its scale factor—by which the distance between galaxies changes. According to T-duality, universes with small scale factors are equivalent to ones with large scale factors. No such symmetry is present in Einstein's equations; it emerges from the unification that string theory embodies, with the dilaton playing a central role.

- For years, string theorists thought that T-duality applied only to closed strings, as opposed to open strings, which have loose ends and thus can't wind. But scientists in the 1990s realized that T-duality did apply to open strings, provided the switch between large and small radii was accompanied by a change in the conditions at the end points of the string. Certain boundary conditions describe how the ends stay put.

- For instance, electrons may be strings whose ends can move around freely in three of the 10 spatial dimensions but are stuck within the other seven. Those three dimensions form a subspace known as a Dirichlet membrane, or D-brane. In 1996, Petr Hořava and Edward Witten proposed that our universe resides on such a brane.

Lesson 2 | The Myth of the Beginning of Time

AN ABHORRENCE OF INFINITY

- The magic properties of quantum strings point in one direction: Strings abhor infinity. They can't collapse to an infinitesimal point, so they avoid the paradoxes that collapse entails. They have nonzero sizes and novel symmetries that set upper bounds to physical quantities that increase without limit in conventional theories, and they set lower bounds to quantities that decrease.

- String theorists expect that when you play the history of the universe backward in time, the curvature of spacetime starts to increase. But instead of going all the way to infinity (at the traditional big bang singularity), it eventually hits a maximum and shrinks again. Before string theory, physicists were hard-pressed to imagine any mechanism that could so cleanly eliminate the singularity.

- Conditions near the zero time of the big bang were so extreme that no one yet knows how to solve the equations. Nevertheless, string theorists have hazarded guesses about the pre-bang universe. Two popular models are floating around.

- The first is known as the pre–big bang scenario, which Gabriele Veneziano and his colleagues began to develop in 1991. It combines T-duality with the better-known symmetry of time reversal, whereby the equations of physics work equally well when applied backward and forward in time.

- The combination gives rise to new possible cosmologies in which the universe, for example, five seconds before the big bang expanded at the same pace as it did five seconds after the bang. Yet the rate of change of the expansion was opposite at the two instants: If it was decelerating after the bang, it was accelerating before.

- The second popular model for the pre-bang universe is the so-called ekpyrotic (or conflagration) scenario. It relies on the idea that our universe sits at one end of a higher-dimensional space and a "hidden brane" sits at the opposite end.

- The two branes exert an attractive force on each other and occasionally collide, making the extra dimension shrink to zero before growing again. The big bang would correspond to the time of collision. In a variant of this scenario, the collisions occur cyclically.

- The pre–big bang and ekpyrotic scenarios share some common features. Both begin with a large, cold, nearly empty universe, and both share the difficult (and unresolved) problem of making the transition between the pre- and the post-bang phase.

ABOUT THIS LESSON

This lesson was adapted from the article "The Myth of the Beginning of Time" by Gabriele Veneziano, one of the founders of string theory and one of the first to later apply the theory to black holes and cosmology.

Lesson 2 Transcript

THE MYTH OF THE BEGINNING OF TIME

This lesson was adapted from an article by Gabriele Veneziano, who is one of the founders of string theory and one of the first to apply string theory to black holes and cosmology.

Was the big bang really the beginning of time? Or did the universe exist before then? Such a question seemed almost blasphemous only decades ago. Most cosmologists insisted that it made no sense—that to contemplate a time before the big bang was like asking for directions to a place north of the North Pole. But developments in theoretical physics, especially the rise of string theory, have changed their perspective. The pre-bang universe has actually become a frontier of cosmology.

In the new willingness to consider what might have happened before the big bang is the latest swing of an intellectual pendulum that's been rocking back and forth for millennia. In one form or another, the issue of the ultimate beginning has engaged philosophers and theologians in nearly every culture. It's entwined with a grand set of concerns, one famously encapsulated in an 1897 painting by Paul Gauguin titled *Where do we come from? What are we? Where are we going?*

The piece depicts the cycle of birth, life and death—origin, identity and destiny for each individual—and these personal concerns connect directly to cosmic ones. We can trace our own lineage back through the generations. In fact, now scientists can trace our lineage through the first land-animal ancestors, to early forms of life, to proto-life, to the elements that were synthesized in the primordial universe, to the amorphous energy deposited in space even before that. Does our family tree extend forever backward? Or do its roots terminate? Is the cosmos as impermanent as we are?

The ancient Greeks debated the origin of time fiercely. Taking the no-beginning side, Aristotle invoked the principle that out of nothing, nothing comes. If the universe could never have gone from nothingness to somethingness, it must always have existed. For this and other reasons, time must stretch eternally into the past and into the future.

Christian theologians tended to take the opposite point of view. Augustine contended that God exists outside of space and time, able to bring these constructs into existence as surely as he would forge other aspects of our world. When asked, "What was God doing before he created the world?" Augustine answered, "Since time itself is part of God's creation, there was simply no before!"

Einstein's general theory of relativity led modern cosmologists to much the same conclusion. The theory holds that space and time are malleable. On the largest scales, space is naturally dynamic, expanding or contracting over time, carrying matter like driftwood on the tide. In the 1920s, astronomers confirmed that our universe is expanding, distant galaxies move apart from one another. One consequence of this, as physicists Stephen Hawking and Roger Penrose proved in the 1960s, is that time cannot extend back indefinitely.

As you play cosmic history backward in time, the galaxies all come together to a single infinitesimal point, known as a singularity—almost as if they were descending into a black hole. Each galaxy or its precursor is squeezed down to zero size. Quantities such as density, temperature and spacetime curvature all become infinite. The singularity is the ultimate cataclysm, beyond which our cosmic ancestry cannot extend.

Lesson 2 Transcript | The Myth of the Beginning of Time

The unavoidable singularity poses serious problems for cosmologists. For the cosmos to look broadly the same everywhere (as it does) some kind of communication had to pass among distant regions of space, somehow coordinating their properties. Yet the idea of such communication contradicts our standard understanding of cosmology.

So, consider what's happened over the past 13.8 billion years since the big bang. The distance between galaxies has grown by a factor of about 1,000 because the universe is expanding, while the radius of the observable universe has grown even more by the factor of about 100,000; and that is because light is even faster than the expansion of the universe.

So today we see parts of the universe that we couldn't have seen 13.8 billion years ago. Indeed, this is the first time in cosmic history that light from the most distant galaxies has reached us in the Milky Way. Nevertheless, the properties of our Milky Way galaxy are basically the same as those of all the distant galaxies. It's like showing up at a party only to find you're wearing exactly the same outfit as a dozen of your friends. If just two of you were dressed the same, it might be explained away as a coincidence, but a dozen suggests that everyone coordinated their attire in advance.

In cosmology, it's not clothes, but independent yet statistically identical patches of sky in what's called the cosmic microwave background radiation, which is the afterglow from the big bang. And the number isn't just a dozen but rather tens of thousands. So, one possible explanation is that all those regions of space were endowed at birth with identical properties—in other words, that the homogeneity is just a coincidence. However, physicists have come up with two other more natural ways out of this impasse: One is that the early universe was either much smaller or much older than is known from standard cosmology. Either (or actually both if they could act together) would have made intercommunication possible.

So, one explanation follows the first alternative. It postulates that the universe went through a period of accelerating expansion, known as inflation, early in its history. Before this phase, galaxies or their precursors were so closely packed that they could easily coordinate their properties; and then, during inflation, they fell out of contact because light couldn't keep pace with the frenetic expansion. After inflation ended, the expansion began to decelerate, so galaxies gradually came back into one another's view.

Physicists ascribe the inflationary spurt to the potential energy stored in a new quantum field, which is called the inflaton, which happened about 10 to the negative-35 second after the big bang. Unlike rest mass or kinetic energy, potential energy leads to gravitational repulsion. So rather than slowing down the expansion, as the gravitation of ordinary matter would, the inflaton accelerated it.

Proposed in 1981, inflation has explained a wide variety of observations with precision. But a number of possible theoretical problems remain, beginning with the questions of what exactly the inflaton was and what gave it such a huge initial potential energy.

So, that's inflation, but another way to solve the puzzle follows the second alternative by just getting rid of the singularity. If time didn't begin at the big bang, and if a long era preceded the onset of the present cosmic expansion, then matter could've had plenty of time to arrange itself smoothly, and this has led some researchers to reexamine the reasoning that led them to infer a singularity in the first place.

So one of the assumptions—that relativity theory is always valid—is questionable. Close to the putative singularity, quantum effects must have been important, or even dominant. Standard relativity takes no account of such effects, so accepting the inevitability of the singularity amounts to trusting the theory really beyond reason. To know what really happened, physicists need to include relativity in a quantum theory of gravity. This task has occupied theorists from Albert Einstein onward, but progress was almost zero until the mid-1980s.

Today two approaches stand out: One is called loop quantum gravity, and it retains Einstein's theory essentially intact but it changes the procedure for implementing it in quantum mechanics. So, practitioners of loop quantum gravity have taken great strides and achieved deep insights over the past several years. Still, their approach may not be enough to resolve the fundamental problems of quantizing gravity.

The second approach is string theory and that is a really revolutionary modification of Einstein's theory, and this is what we'll focus on here, although proponents of loop quantum gravity reach some of the same conclusions. String theory grew out of a model that physicist Gabriele Veneziano wrote in 1968 to describe the world of nuclear particles (which are protons and neutrons) and their interactions. Despite much initial excitement, the model was abandoned several years later in favor of another model called quantum chromodynamics, which describes nuclear particles in terms of more elementary constituents called quarks.

Quarks are confined inside a proton or a neutron as if they were tied together by elastic strings. In retrospect, the original string theory had captured these stringy aspects of the nuclear world, only later it was revived as a candidate for combining general relativity and quantum theory. The basic idea is that elementary particles are not point-like but rather infinitely thin, one-dimensional objects, and these are the strings. So, the zoo we know of elementary particles, each of which its own characteristic properties, those reflect the many possible vibration patterns of a string.

So, how can such a "simple-minded theory," which is what Veneziano calls it, how can that describe the complicated world of particles and all their interactions? And so the answer can be found in what he dubbed "quantum string magic." Once the rules of quantum mechanics are applied to a vibrating string—just like a miniature violin string, except that the vibrations propagate along it at the speed of light—once that happens new properties appear, and all of these have profound implications for particle physics and for cosmology—so the smallest and the biggest.

First, quantum strings have a finite size. If it weren't for quantum effects, a violin string could be cut in half, cut in half again, and so on, all the way down, finally becoming a massless, point-like particle. But Heisenberg's uncertainty principle eventually intrudes if you get small enough, and that prevents even the lightest strings from being sliced smaller than about 10 to the negative–34th meter. This irreducible quantum of length, which is denoted as l sub-s, is a new constant of nature introduced by string theory. It plays a crucial role in almost every aspect of string theory, and it puts a finite limit on quantities that otherwise could become either zero or infinite.

So, second, quantum strings may have angular momentum even if they lack mass. Now, in classical physics, angular momentum is a property of an object that rotates with respect to an axis; so, the formula for angular momentum multiplies together velocity, mass and distance from the axis; so, hence, a massless object can't have angular momentum. But quantum fluctuations change the situation. A tiny string can acquire a certain amount of angular momentum even without having any mass. So, this feature precisely matches the properties of all the fundamental forces, such as the photon (which is for electromagnetism) and the graviton (for gravity). Historically, angular momentum is what clued physicists into the quantum-gravitational implications of string theory.

So, third, quantum strings demand the existence of extra dimensions of space. Whereas a classical violin string will vibrate no matter what the properties of space and time are, a quantum string is more finicky. The equations describing the vibration become inconsistent unless spacetime is either highly curved (which is contrary to what we observe) or if it contains six additional spatial dimensions.

Fourth, physical constants no longer have arbitrary, fixed values. Instead, they occur in string theory as fields, rather like the electromagnetic field, that can adjust their values dynamically. So, these fields may have taken different values in different cosmological epochs or in remote regions of space, and even today the physical "constants" that we know may vary by a small amount and observing any variation would provide an enormous boost to string theory.

One of the fields, called the dilaton, is the master key to string theory; it determines the overall strength of all interactions. The dilaton fascinates string theorists because its value can be reinterpreted as the size of an extra dimension of space, giving a grand total of 11 spacetime dimensions.

So, finally, quantum strings have introduced physicists to some striking new symmetries of nature known as dualities, which alter our intuition for what happens when objects get extremely small. Typically, a short string is lighter than a long one, but if we attempt to squeeze down its size below the fundamental length of l sub-s, then the string gets heavier again.

So, another form of symmetry is called T-duality and it holds that small and large extra dimensions are equivalent, and this symmetry arises because strings can move in more complicated ways than point-like particles can. So, consider a closed string (which is a loop) located on a cylindrically shaped space, whose circular cross section represents one finite extra dimension. So, besides vibrating, the string can either turn as a whole around the cylinder or wind around it, one or several times, like a rubber band wrapped-up around a rolled-up poster.

So, the energetic cost of these two states of the string depends on the size of the cylinder. The energy of winding is directly proportional to the cylinder radius; larger cylinders require the string to stretch more as it wraps around and the windings contain more energy than they would on a smaller cylinder. The energy associated with moving around the circle, on the other hand, is inversely proportional to the radius: larger cylinders allow for longer wavelengths (which have smaller frequencies), and that represents less energy than shorter wave lengths do.

So, if a large cylinder is substituted for a small one, the two states of motion can swap roles. Energies that had been produced by circular motion are instead produced by winding, and vice versa. An outside observer notices only the energy levels, not the origin of those levels; so, to the observer, the large and small radii are physically equivalent.

So, although this T-duality is usually described in terms of cylindrical spaces, where one dimension is finite, a variant of it applies also to our ordinary three dimensions, which appear to stretch on indefinitely. But we have to be careful when talking about the expansion of an infinite space. Its overall size can't change; it remains infinite. But it can still expand in the sense that bodies embedded within it, such as galaxies, can keep moving apart from one another.

The crucial variable is not the size of the space as a whole but what is called its scale factor and the scale factor describes how the distance between galaxies changes. According to T-duality, universes with small scale factors are equivalent to ones with large scale factors. But no such symmetry is present in Einstein's equations; it emerges from the unification that string theory embodies, with the dilaton playing a central role.

For years string theorists thought that T-duality applied only to closed strings, as opposed to open strings, which have loose ends and thus can't wind. But in the 1990s, scientists realized that T-duality did apply to open strings, provided the switch between the large and small radii was accompanied by a change in the conditions at the end points of the string and certain boundary conditions describe how those ends stay put.

For instance, electrons may be strings whose ends can move around freely in three of the 10 spatial dimensions but are stuck within the other seven. Those three dimensions form a subspace known as a Dirichlet membrane, or

D-brane. In 1996 Petr Hořava of the University of California, Berkeley, and Edward Witten of the Institute for Advanced Study in Princeton, New Jersey, proposed that our universe resides on such a brane. The partial mobility of electrons and other particles explains why we're unable to perceive the full 10-dimensional glory of space.

All the magic properties of quantum strings point in one direction: strings abhor infinity. They can't collapse to an infinitesimal point, so they avoid the paradoxes that collapse entails. Their nonzero size and their novel symmetries set upper bounds to the physical quantities that increase without limit in conventional theories and they set lower bounds to quantities that decrease.

String theorists expect that when you play the history of the universe backward in time, the curvature of spacetime starts to increase. But instead of going all the way to infinity (which is the traditional big bang singularity), it eventually hits a maximum and then shrinks again. Before string theory, physicists were hard-pressed to imagine any mechanism that could so clearly eliminate the need for a singularity.

Conditions near the zero time of the big bang were so extreme that no one yet knows how to solve the equations. Nevertheless, string theorists have hazarded guesses about what's called the pre-bang universe and two popular models are still floating around. The first is known as the pre-big bang scenario, which Gabriele Veneziano and his colleagues began to develop in 1991. It combines T-duality with the better-known symmetry of time reversal, whereby the equations of physics work equally well when applied backward and forward in time.

The combination gives rise to new possible cosmologies in which the universe, say, five seconds before the big bang expanded at the same pace as it did five seconds after the bang. Yet the rate of change of the expansion was opposite at the two instances: If it was decelerating after the bang, it was accelerating before. So, in short, the big bang may not have been the origin of the universe but instead simply a violent transition from acceleration to deceleration.

The beauty of this picture is that it automatically incorporates the great insight of standard inflationary theory—namely, that the universe had to undergo a period of acceleration to become so homogeneous. In the standard theory, acceleration occurs after the big bang because of an ad hoc inflaton field. In the pre-big bang scenario, it happens before the bang as a natural consequence of the novel symmetries of string theory.

According to this scenario, the pre-bang universe was almost a perfect mirror image of the post-bang. If the universe is eternal into the future, it's also eternal into the past. Infinitely long ago it was nearly empty, filled only with a tenuous, widely dispersed, chaotic gas of radiation and matter. The forces of nature, controlled by the dilaton field, were so feeble that particles in this gas barely interacted.

As time went on, the forces gained in strength and pulled matter together. Randomly, some regions accumulated matter at the expense of their surroundings and eventually the density in these regions became so high that black holes started to form. Matter inside those regions was then cut off from the outside, breaking the universe up into disconnected pieces.

Inside a black hole, space and time swap roles. The center of the black hole isn't a point in space but an instant in time. As the infalling matter approached the center, it reached higher and higher densities. But when the density, temperature and curvature reached the maximum values allowed by string theory, these quantities bounced and started decreasing. The moment of that reversal, called a big bang, was later renamed a big bounce. The interior of one of these black holes became our universe.

Not surprisingly, such an unconventional scenario has provoked a lot of controversy. Andrei Linde of Stanford University has argued that for this scenario to match observations, the black hole that gave rise to our universe would've needed to be unusually large—much larger than the length scale of string theory. So, an answer to this objection is that the equations predict black holes of all possible sizes. Our universe just happened to form inside a sufficiently large one.

So, a more serious objection argues that matter and spacetime would've behaved chaotically near the moment of the bang, in possible contradiction with the observed regularity of the early universe. Gabriele Veneziano has proposed that a chaotic state would produce a dense gas of miniature string holes—strings that were so small and massive that they were on the verge of becoming black holes.

The second popular model for the pre-bang universe is the so-called ekpyrotic (or conflagration) scenario. It relies on the idea that our universe sits at one end of a higher-dimensional space and a hidden brane sits at the opposite end. The two branes exert an attractive force on each other and occasionally collide, making the extra dimension shrink to zero before growing again. The big bang would correspond to the time of collision.

In a variant of this scenario, the collisions occur cyclically. Two branes might hit, bounce off each other, move apart, pull together again, hit again, and so on. In between collisions, the branes behave like Silly Putty, expanding as they recede and contracting somewhat as they come back together. During the turnaround, the expansion rate accelerates; indeed, the present accelerating expansion of our universe may augur another collision.

So, the pre-big bang and ekpyrotic scenarios share some common features. Both begin with a large, cold, nearly empty universe, and both share the difficult problem of making the transition between the pre- and the post-bang phase. Mathematically, the main difference between the scenarios is the behavior of the dilaton field. In the pre-big bang, the dilaton begins with a low value so that the forces of nature are weak and then steadily gains strength. The opposite is true for the ekpyrotic scenario, in which the collision occurs when forces are at their weakest.

The developers of the ekpyrotic theory initially hoped the weakness of the forces would allow the bounce to be analyzed more easily, but they were still confronted with a difficult high-curvature situation. So, the jury's still out on whether the scenario truly avoids a singularity.

Also, the ekpyrotic scenario must entail very special conditions to solve some of the usual cosmological puzzles. For instance, the branes when they are about to collide must've been almost exactly parallel to one another, otherwise the collision wouldn't have given rise to a sufficiently homogeneous bang. The cyclic version may be able to take care of this problem, because successive collisions would have allowed the branes to straighten themselves.

Leaving aside the difficult task of fully justifying these two scenarios mathematically, physicists must also ask whether they have any observable physical consequences. For at first sight, both scenarios might seem like an exercise not in physics but in metaphysics—interesting ideas that observers could never prove either right or wrong. But a possible pre-bangian epoch could have observable consequences, especially for the very small variations observed in the cosmic microwave background temperature.

First, observations show that the temperature fluctuations were shaped by acoustic waves for several hundred thousand years. The regularity of the fluctuations indicates that the waves were synchronized. Cosmologists have discarded many models over the years because they failed to account for this synchrony. But the inflationary, pre-big bang and ekpyrotic scenarios all pass this first test. In these models, the waves were triggered by quantum processes amplified during the period of accelerating cosmic expansion. The phases of these waves were aligned.

Second, each model predicts a different distribution of the temperature fluctuations with respect to angular size. Observers have found that fluctuations of all sizes have approximately the same amplitude. So, discernible deviations occur only on very small scales, and for that primordial fluctuations have been altered by subsequent processes, so we don't have to worry about that as much. Inflationary models neatly reproduce this distribution. During inflation, the curvature of space changed relatively slowly, so fluctuations of different sizes were generated under much the same conditions.

In both the stringy models, the curvature evolved quickly, increasing the amplitude of small-scale fluctuations, but other processes boosted the large-scale ones, leaving all fluctuations with the same strength. For the ekpyrotic scenario, those other processes involved the extra dimension of space, the one that separated the colliding branes. For the pre-big bang scenario, they involved a quantum field called the axion, related to the dilaton. In short, all three models do match the data.

Variations in temperature can arise from two distinct processes in the early universe: Fluctuations in the density of matter and rippling caused by gravitational waves. Inflation involves both of these processes, whereas the pre-big bang and ekpyrotic scenarios involve density variations for the most part. Gravitational waves of certain sizes would leave a distinctive signature in the polarization of the cosmic microwave background radiation. Satellite and ground-based observations may be able to see that signature, if it exists. That would provide a really definitive test.

A fourth test pertains to the statistics of the fluctuations. In inflation, the fluctuations follow a bell-shaped curve and the same may be true in the ekpyrotic case, whereas the pre-big bang scenario allows for sizable deviation from such a curve. But analysis of the cosmic microwave background radiation isn't the only way to verify these theories. The pre-big bang scenario should also produce a random background of gravitational waves in a range of frequencies that, though they're irrelevant for the microwave background, should be detectable by future gravitational-wave observatories.

Moreover, because the pre-big bang universe and the ekpyrotic scenarios involve changes in the dilaton field, which is coupled to the electromagnetic field, they would both lead to large-scale magnetic field fluctuations.

Vestiges of these fluctuations might show up in galactic and intergalactic magnetic fields.

So, when did time begin? Science doesn't have a conclusive answer yet, but at least two potentially testable theories plausibly hold that the universe—and therefore time—existed well before the big bang. If either scenario is right, that means the cosmos has always existed and, even if it collapses again one day, it always will exist.

3
THAT MYSTERIOUS FLOW

This lesson looks at the passage of time, which is probably the most basic facet of human perception: We feel time slipping by in our innermost selves in a manner that is far more intimate than our experience of, for instance, space or mass. However, nothing in known physics corresponds to the passage of time. Physicists insist that time doesn't flow at all; it merely is. Some philosophers argue that the very notion of the passage of time is nonsensical.

Lesson 3 | That Mysterious Flow

PAST, PRESENT, AND FUTURE

- In daily life, we divide time into three parts: past, present and future. Reality is associated with the present moment. The past we think of as having slipped out of existence, whereas the future is even more shadowy, its details still unformed.

- In this simple picture, the "now" of our conscious awareness glides steadily onward, transforming events that were once in the unformed future into the concrete-but-fleeting reality of the present—and then relegating them to the fixed past. However, this is at odds with modern physics. Albert Einstein famously expressed this point when he wrote to a friend, "The past, present, and future are only illusions, even if stubborn ones."

- Einstein's startling conclusion stems from his special theory of relativity, which denies any absolute, universal significance to the present moment. According to the theory, simultaneity is relative. Two events that occur at the same moment if observed from one reference frame may occur at different moments if viewed from another.

"The past, present, and future are only illusions, even if stubborn ones."

INFERRING ANSWERS

- An innocuous question such as "What's happening on Mars now?" has no definite answer. The key point is that Earth and Mars are a long way apart—up to about 20 light-minutes.

- Because information can't travel faster than light, an Earth-based observer is unable to know the situation on Mars at the same instant. They must infer the answer after the event, when light has had a chance to pass between the planets. The inferred past event will be different depending on the observer's velocity.

- For example, during a future crewed expedition to Mars, mission controllers back on Earth might say, "I wonder what Commander Jones is doing at Alpha Base right now." The trouble stems from the phrase "right now." Different people who are moving at different velocities have different perceptions of what the present moment is. This strange fact is known as the relativity of simultaneity.

- In the following scenario, two people—an earthling sitting in Houston and an astronaut crossing the solar system from Earth to Mars at 80% of the speed of light—attempt to answer the question of what's happening with a third person, Commander Jones, on Mars right now. On Mars, Commander Jones at Alpha Base has agreed to eat lunch when her clock strikes 12:00 pm and to transmit a signal at the same time.

  - From the earthling's perspective, Earth is standing still, Mars is a constant distance (20 light-minutes) away, and the rocket ship is moving at 80% of the speed of light. The situation looks exactly the same to the martian, Commander Jones.

  - Before noon, by exchanging light signals, the earthling and martian measure the distance between them and synchronize their clocks.

- At 12:00 pm, the earthling hypothesizes that the martian has begun to eat her lunch. He prepares to wait 20 minutes for verification.

- At 12:11 pm, knowing the rocket's speed, the earthling deduces that it encounters the signal while on its way to Mars.

- At 12:20 pm, the signal arrives at Earth. The earthling has confirmed his earlier hypothesis. Noon on Mars is the same as noon on Earth.

- At 12:25 pm, the ship arrives at Mars.

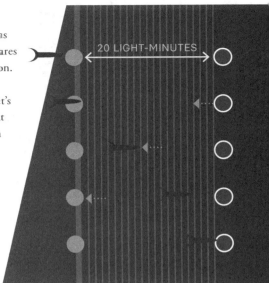

However, the situation as seen from the rocket is different. From the rocket man's perspective, the rocket is standing still. It is the planets that are hurtling through space at 80% of the speed of light.

His measurements show the two planets to be separated by 12 light-minutes—a different distance from what the earthling inferred. This discrepancy, an effect of Einstein's theory, is called length contraction.

A related effect, time dilation, causes clocks on the ship and planets to run at different rates. As the ship passes Earth, it synchronizes its clock to Earth's. Before noon, by exchanging light signals with his colleagues, the rocket man measures the distance between the planets.

- At 12:00 pm, while passing Earth, the rocket man hypothesizes that Commander Jones has begun to eat. He prepares to wait 12 minutes for verification.

▼ At 12:07 pm, the signal arrives, disproving the hypothesis. The rocket man infers that the Martian ate sometime before noon (rocket time).

▼ At 12:15 pm, Mars arrives at the ship. The rocket man and Commander Jones notice that their two clocks are out of sync but disagree as to whose is right.

▼ At 12:33 pm, the signal arrives at Earth. The clock discrepancies demonstrate that there is no universal present moment.

◢ Such mismatches make a mockery of any attempt to confer special status on the present moment, for whose "now" does that moment refer to? If two people are in relative motion, an event that one might judge to be in the as-yet-undecided future might already exist in the fixed past for the other.

◢ The most straightforward conclusion is that both past and future are fixed. For this reason, physicists prefer to think of time as laid out in its entirety—a timescape, analogous to a landscape—with all past and future events located there together. It's a notion sometimes referred to as block time. Under this line of thought, the time of the physicist does not pass or flow.

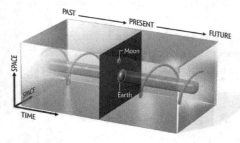

Conventional view: Only the present is real

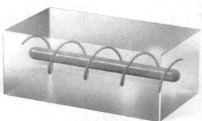

Block universe: All times are equally real

Lesson 3 | That Mysterious Flow

PHILOSOPHERS' CONCLUSIONS

- A number of philosophers over the years have arrived at the same conclusion by examining what we normally mean by the passage of time. They argue that the notion is internally inconsistent. The concept of flux, after all, refers to motion.

- It makes sense to talk about the movement of a physical object, such as an arrow through space, by gauging how its location varies with time. But what meaning can be attached to the movement of time itself?

- Although we find it convenient to refer to time's passage in everyday affairs, the notion imparts no new information that can't be conveyed without it. Consider the following description:

    > Alice was hoping for a white Christmas, but when the day came, she was disappointed that it only rained. However, she was happy that it snowed the following day.

- Although that description is replete with tenses and references to time's passage, exactly the same information is conveyed by simply correlating Alice's mental states with dates. Thus, the following cumbersome and rather dry catalogue of facts suffices:

    > December 24: Alice hopes for a white Christmas.
    > December 25: It rains. Alice is disappointed.
    > December 26: It snows. Alice is happy.

- In this description, nothing happens or changes. There are simply states of the world at different dates and associated mental states for Alice. Similar arguments go back to ancient Greek philosophers such as Parmenides and Zeno.

- A century ago, the British philosopher John McTaggart sought to draw a clear distinction between the description of the world in terms of events happening, which he called the A series, and the description in terms of dates correlated with states of the world, the B series. Each appears to be a true description of reality, and yet the two points of view are seemingly in contradiction.

- For example, the event "Alice is disappointed" was once in the future, then in the present, and afterward in the past. But past, present, and future are exclusive categories, so how can a single event have the character of belonging to all three?

- McTaggart used this clash between the A and B series to argue for the unreality of time as such. Most physicists would put it less dramatically: The flow of time is unreal, but time itself is as real as space.

TIME ASYMMETRY

- A great source of confusion in discussions of time's passage stems from its link with the so-called arrow of time. To deny that time flows is not to claim that the designations *past* and *future* are without physical basis. Events in the world undeniably form a unidirectional sequence.

- Imagine an egg dropped on the floor. It will smash into pieces. The reverse process—a broken egg spontaneously assembling itself into an intact egg—is something we never see.

- This is an example of the second law of thermodynamics, which states that the entropy of a closed system—roughly defined as how disordered it is—will tend to rise with time. An intact egg has lower entropy than a shattered one.

Lesson 3 | That Mysterious Flow

- Because nature abounds with irreversible physical processes, the second law of thermodynamics plays a key role in imprinting on the world a conspicuous asymmetry between past and future directions along the time axis. By convention, the arrow of time points toward the future.

- This does not imply, though, that the arrow is moving toward the future any more than a compass needle pointing north indicates that the compass is traveling north. Both arrows symbolize an asymmetry, not a movement.

- Some researchers have contended that the subtle physics of irreversible processes make the flow of time an objective aspect of the world. But theoretical physicist Paul Davies and others argue that it's really just an illusion.

- After all, we don't really observe the passage of time. In reality, we observe that later states of the world differ from earlier states we still remember. The fact that we remember the past, rather than the future, is an observation not of the passage of time but of the asymmetry of time.

- Only conscious observers can register the flow of time. Much as a measuring tape measures distances between places, a clock measures durations between events. However, that is all it measures. It doesn't measure the speed with which one moment succeeds another. It appears the flow of time is subjective, not objective. This illusion cries out for explanation, and that explanation is to be sought in psychology, neurophysiology, and perhaps linguistics or culture.

PERCEPTIONS OF TIME

- Since modern science has barely begun to consider the question of how we perceive the passage of time, we can only speculate about the answer. It might have something to do with the functioning of the brain. There are two aspects to time asymmetry that might create the false impression that time is flowing.

- The first is the thermodynamic distinction between past and future. As physicists have come to realize, the concept of entropy is closely related to the information content of a system. For this reason, the formation of memory is a unidirectional process: New memories add information and raise the entropy of the brain. We might perceive this unidirectionality as the flow of time.

- A second possibility is that our perception of the flow of time is linked in some way to quantum mechanics. From the earliest days of the formulation of quantum mechanics, physicists recognized that time enters into the theory in a unique manner, quite unlike space. The special role of time is one reason it is proving so difficult to merge quantum mechanics with general relativity.

- The Heisenberg uncertainty principle, which says that nature is inherently indeterministic, implies an open future (and, for that matter, an open past). This indeterminism manifests itself most conspicuously on an atomic scale of size and dictates that the observable properties that characterize a physical system are generally undecided from one moment to the next.

- For example, an electron hitting an atom may bounce off in one of many directions, and it is normally impossible to predict in advance what the outcome in any given case will be. Quantum indeterminism implies that for a particular quantum state, there are many (possibly infinite) alternative futures or potential realities. Quantum mechanics supplies the relative probabilities for each observable outcome, although it won't say which potential future is destined for reality.

- However, when a human observer makes a measurement, only one result is obtained; for example, the rebounding electron will be found moving in a certain direction. In the act of measurement, a single, specific reality becomes projected out from a vast array of possibilities.

- Within the observer's mind, the possible makes a transition to the actual. It transitions from the open future to the fixed past—which is precisely what we mean by the flux of time.

- Unfortunately, physicists disagree on how this transition from many potential realities into a single actuality takes place. Many have argued it has something to do with the consciousness of the observer—that it's the act of observation itself that prompts nature to make up its mind. A few researchers maintain that consciousness—including the impression of temporal flux—could be related to quantum processes that take place in the brain.

ABOUT THIS LESSON

This lesson was adapted from the article "That Mysterious Flow" by Paul Davies, a theoretical physicist and cosmologist at Arizona State University, where he is director of the Beyond Center for Fundamental Concepts in Science.

Lesson 3 Transcript

THAT MYSTERIOUS FLOW

This lesson is adapted from an article by Paul Davies, a theoretical physicist and cosmologist at Arizona State University, where he is director of BEYOND: The Center for Fundamental Concepts in Science.

Augustine of Hippo, the 5th-century theologian, remarked that he knew well what time is—until somebody asked. Then he was at a loss for words. Because we sense time psychologically, definitions of time based on physics seem dry and inadequate. For the physicist, time is simply what clocks measure. Mathematically, it's a one-dimensional space, usually assumed to be continuous, although it might be quantized into discrete chronons, like the frames of a movie.

To be perfectly honest, neither scientists nor philosophers really know what time is or why it exists. The best thing they can say is that time is an extra dimension akin to space. For example, the two-dimensional orbit of the moon through space can be thought of as a three-dimensional corkscrew through spacetime. The fact that time can be treated as a fourth dimension does not mean it's identical to the three dimensions of space. Time and space enter into daily experience and physical theory in distinct ways. For instance, the distinction between space and time underpins the key notion of causality, stopping cause and effect from being just hopelessly jumbled.

On the other hand, many physicists believe that on the very smallest scale of size and duration, space and time might lose their separate identities.

"Gather ye rosebuds while ye may, / Old Time is still a-flying." So wrote 17th-century English poet Robert Herrick, capturing the universal experience that time flies. The passage of time is probably the most basic facet of human perception, for we feel time slipping by in our innermost selves in a manner that's far more intimate than our experience of, say, space or mass. The passage of time has been compared to the flight of an arrow and to an ever-rolling stream, bearing us inexorably from past to future. Shakespeare wrote of "the whirligig of time," and his countryman Andrew Marvell of "Time's wingèd chariot hurrying near."

Now, as evocatively, they run afoul of a deep and devastating paradox. Nothing in known physics corresponds to the passage of time. Indeed, physicists insist that time doesn't flow at all; it merely is. Some philosophers argue that the very notion of the passage of time is nonsensical and that talk of the river or flux of time is founded on a misconception. So, how can something so basic to our experience of the physical world turn out to be a case of mistaken identity? Is there a key quality of time that science has not yet identified?

In daily life we divide time into three parts: past, present and future. Reality is associated with the present moment. The past we think of as having slipped out of existence, whereas the future is even more shadowy, its details still unformed. In this simple picture, the *now* of our conscious awareness glides steadily onward, transforming events that were once in the unformed future into the concrete but fleeting reality of the present and then relegating them to the fixed past.

Though this commonsense description may seem obvious, it's seriously at odds with modern physics. Albert Einstein famously expressed this point when he wrote to a friend, "The past, present and future are only illusions, even if stubborn ones." Einstein's startling conclusion stems directly from his special theory of relativity, which denies any absolute, universal significance to the present moment. According to the theory, simultaneity is relative. Two events that occur at the same time if observed from one reference frame may occur at different moments if viewed from another.

An innocuous question such as "What's happening on Mars right now?" has no definite answer. The key point is that Earth and Mars are a long way apart—up to about 20 light-minutes. Because information can't travel faster than light, an Earth-based observer is unable to know the situation on Mars at the same instant. They must infer the answer after the event, when light has had a chance to pass between the planets. The inferred past event will be different depending on the observer's velocity.

So, for example, during a future crewed mission to Mars, controllers back on Earth might say, "I wonder what Commander Jones is doing at Alpha Base right now." What's happening on Mars right now? Such a simple question, such a complex answer. The trouble stems from the phrase "right now." Different people, moving at different velocities, have different perceptions of what the present moment is. This strange fact is known as the relativity of simultaneity.

In the following scenario, two people—an earthling sitting in Houston and an astronaut crossing the solar system from Earth to Mars at say 80% of the speed of light—attempt to answer the question of what's happening with a third person who is on Mars right now. On Mars, commander Jones at Alpha Base has agreed to eat lunch when her clock strikes 12:00 noon and to transmit a signal at that time.

Now, from the earthling's perspective, Earth is standing still, Mars is a constant distance away, about 20 light-minutes, and the rocket ship is moving at 80% of the speed of light. The situation looks exactly the same to the martian, Commander Jones, who is about to eat lunch. Before noon, by exchanging light signals, the earthling and martian measure the distance between them and synchronize their clocks.

At 12:00 noon, the earthling hypothesizes that the martian has begun to eat her lunch. He prepares to wait 20 minutes for verification. At 12:11 pm, knowing the rocket's speed, the earthling deduces that the rocket is encountering the signal while it is on its way to Mars. At 12:20 pm, the signal from Mars arrives at Earth. The earthling has confirmed his earlier hypothesis. Noon on Mars is the same as noon on Earth. At 12:25 pm, the ship arrives at Mars.

Now let's view the situation as seen from the rocket ship. From the rocket man's perspective, the rocket is standing still. It's the planets that are hurtling through space at 80% of the speed of light. His measurements show the two planets to be separated by just 12 light-minutes—a different distance from what the earthling inferred. This discrepancy, which is an effect of Einstein's theory, is called length contraction.

A related effect, called time dilation, causes clocks on the ship and the planets to run at different rates. As the ship passes Earth, it synchronizes its clock to Earth's clock. Before noon, by exchanging light signals with his colleagues, the rocket man measures the distance between the planets. At 12:00 noon, while passing the Earth, the rocket man hypothesizes that Commander Jones has begun to eat. He prepares to wait 12 minutes for verification. At 12:07 pm, the signal arrives, disproving the hypothesis. The rocket man infers that the martian ate sometime before noon, according to rocket time. At 12:15 pm, Mars arrives at the ship. The rocket man and Commander Jones notice that their two clocks are out of sync but disagree as to whose clock is right. At 12:33 pm, the signal arrives at Earth. The clock discrepancies demonstrate that there is no universal present moment.

Such mismatches make a mockery of any attempt to confer some special status on the present moment, for whose *now* does that moment refer to? If you and I were in relative motion, an event that I might judge to be in the yet undecided future might for you already exist in the fixed past.

The most straightforward conclusion is that both past and future are fixed. For this reason, physicists prefer to think of time as laid out in its entirety—a timescape, analogous to a landscape—with all past and future events located there together. It's a notion sometimes referred to as block time. Completely absent from this description of nature is anything that singles out a privileged, special moment as the present or any process that would systematically turn future events into the present and then the past. In short, the time of the physicist does not pass or flow. So, time doesn't really fly after all.

A number of philosophers over the years have arrived at the same conclusion by examining what we normally mean by the passage of time. They argue that the notion is internally inconsistent. The concept of flux, after all, refers to motion. It makes sense to talk about the movement of a physical object, such as an arrow through space, by gauging how its location varies according to time. But what if meaning can be attached to the movement of time itself? Relative to what does time move? Whereas other types of motion relate one physical process to another, the putative flow of time relates time to itself. Posing the simple question "How fast does time pass?" exposes the absurdity of the very idea. The answer is "one second per second" and that tells us nothing at all.

So, although we find it convenient to refer to time's passage in everyday affairs, the notion imparts no new information that can't be conveyed without it. Consider the following scenario: Alice was hoping for a white Christmas, but when the day came, she was disappointed that it only rained. However, she was happy that it snowed the following day. Although this description is replete with verb tenses and references to time's passage, exactly the same information is conveyed by simply correlating Alice's mental states with dates, all in a way that omits reference to time passing or the world changing.

So, thus, the following cumbersome and rather dry catalogue of facts suffices; December 24: Alice hopes for a white Christmas; December 25: It rains. Alice is disappointed; December 26: It snows. Alice is happy. In this description, nothing happens or and nothing changes. There are simply states of the world at different dates and associated with different mental states for Alice.

Similar arguments go back to ancient Greek philosophers such as Parmenides and Zeno. And a century ago, British philosopher John McTaggart sought to draw a clear distinction between the description of the world in terms of events happening, which he called the A series, and the description in terms of dates correlated with states of the world, or what he called the B series. Each appears to be a true description of reality, and yet the two points of view are seemingly in contradiction.

For example, the event "Alice is disappointed" was once in the future, then it was in the present and afterward in the past. But past, present and future are exclusive categories, so how can a single event have the character of belonging to all three? McTaggart used this clash between the A and B series to argue for the unreality of time as such; which was perhaps a rather drastic conclusion. Most physicists would put it less dramatically: The flow of time is unreal, but time itself is as real as space.

A great source of confusion in discussions of time's passage stems from its link with the so-called arrow of time. To deny that time flows is not to claim that the designations of past and future are without physical basis. Events in the world undeniably form a unidirectional sequence.

So, take an egg dropped on the floor. It'll smash into pieces. Whereas the reverse process—a broken splattered egg spontaneously assembling itself into an intact egg—is something we never see. This is an example of the second law of thermodynamics, which states that the entropy of a closed system— which is roughly defined as how disordered it is—will tend to rise with time. An intact egg has lower entropy than a shattered one. Because nature abounds with irreversible physical processes, the second law of thermodynamics plays a key role in imprinting on the world a conspicuous asymmetry between the past and the future directions along the time axis.

Now, by convention, the arrow of time points toward the future. This does not imply that the arrow is moving toward the future, any more than a compass needle pointing north indicates that the compass is traveling north. Both arrows symbolize an asymmetry, but not a movement. The arrow of time denotes an asymmetry of the world in time, not an asymmetry or flux of time. So, the labels *past* and *future* may legitimately be applied to temporal directions, just as *up* and *down* may be applied to spatial directions but talk of the past or the future is as meaningless as referring to the up or the down.

The distinction between the pastness or futureness, or the past or the future, is graphically illustrated by imagining, say, a movie of an egg being dropped on the floor and splattering. If the film were run backward through the projector, everyone would know that that sequence was unreal. But now imagine that the film's strip has been cut up into frames and the frames have been shuffled randomly. It would be a pretty straightforward task for someone to rearrange the stack of frames into a correctly ordered sequence, with the broken egg at the top and the intact egg at the bottom of the stack.

This vertical stack retains the asymmetry implied by the arrow of time because it forms an ordered sequence in vertical space, proving that time's asymmetry is actually a property of states of the world, not a property of time as such. It's not necessary for the film to actually be run as a movie for the arrow of time to be discerned. Given that most physical and philosophical analyses of time fail to uncover any sign of temporal flow, we're left with something of a mystery. To what should we attribute the powerful, universal impression that the world's in a continual state of flux?

Some researchers have contended that the subtle physics of irreversible processes make the flow of time an objective aspect of the world. But theoretical physicist Paul Davies and others argue that it's really just an illusion. After all, we don't really observe the passage of time. What we actually observe is that later states of the world differ from earlier states; the ones we still remember. The fact that we remember the past, rather than remembering the future, is an observation not of the passage of time but of the asymmetry of time.

Only conscious observers can register the flow of time. Much as a measuring tape measures distances between places, a clock measures durations between events. But that's all it measures. It doesn't measure the speed with which one moment succeeds another. And so, it appears the flow of time is subjective, not objective. This illusion cries out for explanation, and that explanation is to be sought in psychology, neurophysiology, and maybe linguistics or culture. Since modern science has barely begun to consider the question of how we perceive the passage of time, we can only speculate about the answer.

It might have something to do with the functioning of the brain. So, imagine, if you spin around several times and stop suddenly, you'll feel dizzy. Subjectively, it seems as if the world were rotating relative to you, but the evidence of your eyes is clear enough, it's not rotating. The apparent movement of your surroundings is just an illusion created by the rotation of fluid in the inner ear. Perhaps temporal flux is something similar.

There are two aspects to time asymmetry that might create the false impression that time is flowing. The first is called thermodynamic distinction between past and future. As physicists have come to realize over the past few decades, the concept of entropy is closely related to the information content of a system. For this reason, the formation of memory is a unidirectional process; new memories add information and raise the entropy of the brain. We might perceive this unidirectionality as the flow of time.

A second possibility is that our perception of the flow of time is linked in some way to quantum mechanics. From the earliest days of the formulation of quantum mechanics, physicists recognized that time enters into the theory in a unique manner, quite unlike space. The special role of time is one reason it's proving so difficult to merge quantum mechanics with general relativity.

The Heisenberg uncertainty principle, which says that nature is inherently indeterministic, implies an open future (and, for that matter, it also implies an open past). This indeterminism manifests itself most conspicuously on an atomic scale of size and dictates that the observable properties that characterize a physical system are generally undecided from one moment to the next.

For example, an electron hitting an atom may bounce off in one of many different directions and it is normally impossible to predict in advance what the outcome in any specific case will be. Quantum indeterminism implies that for a particular quantum state there are many (possibly infinite) alternative futures or potential realities. Quantum mechanics supplies the relative probabilities for each observable outcome, although it won't say which potential future is destined to become reality.

But when a human observer makes a measurement, one and only one result is obtained; for example, the rebounding electron will be found moving in a certain direction. In the act of measurement, a single, specific reality gets projected out from a vast array of possibilities. Within the observer's mind, the possible makes a transition to the actual, the open future to the fixed past, and that is precisely what we mean by the flux of time.

Now, unfortunately, physicists disagree on how this transition from many potential realities to a single actuality takes place. Many have argued it has something to do with the consciousness of the observer—that it's the act of observation itself that prompts nature to make up its mind. A few researchers maintain that consciousness—including the impression of temporal flux—could be related to quantum processes that take place in the brain.

Although researchers have failed to find evidence for a single "time organ" in the brain—similar to, say, the visual cortex that processes vision—perhaps future work will pin down those brain processes responsible for our sense of temporal passage. It's even possible to imagine there might be drugs that could suspend the subject's impression of time passing. Indeed, some practitioners of meditation already claim to be able to achieve such mental states.

And what if science were able to explain away the flow of time? Perhaps we'd no longer fret about the future or grieve for the past. Worries about death might become as irrelevant as worries about birth. Expectation and nostalgia might cease to be part of human vocabulary. Above all, the sense of urgency that attaches to so much of human activity might evaporate. No longer would we be captives to Henry Wadsworth Longfellow's entreaty to "act, act in the living present," for the past, present and future would literally be things of the past.

4
IS TIME AN ILLUSION?

This lesson considers the question posed by its title: Is time an illusion? To us, it feels like time flows. However, as natural as this way of thinking is, it isn't reflected in science. Physics equations are like a map without the "You Are Here" symbol. The present moment doesn't exist in them, and therefore neither does the flow of time. In fact, Einstein's theories of relativity suggest not only that there is no single special present but also that all moments are equally real. The future is no more open than the past.

EMERGENT TIME

- The gap between our scientific understanding of time and our everyday experience has troubled thinkers throughout history. It has widened as physicists have gradually stripped time of most of the attributes we commonly ascribe to time. This rift is reaching its logical conclusion, for many theoretical physicists have come to believe that time fundamentally does not exist.

- A timeless theory faces the challenge of explaining how we see change if the world isn't really changing. Recent research attempts to pull off just this feat. Although time may not exist at a fundamental level, it may arise at higher levels—just as a table feels solid even though it is actually a swarm of particles composed mostly of empty space. Solidity is a collective, or emergent, property of the particles. Time, too, could be an emergent property of whatever the basic ingredients of the world are.

- This concept of emergent time seems as revolutionary as the theories of relativity and quantum mechanics a century ago. Einstein said that the key step forward in developing relativity was his reconceptualization of time. As physicists pursue his dream of unifying relativity with quantum mechanics, they believe time is again central.

Lesson 4 | Is Time an Illusion?

NEWTONIAN TIME

- Newton's laws of motion require time to have many specific features. No matter when or where an event occurs, classical physics assumes all observers can objectively say whether it happens before, after, or simultaneously with any other event in the universe.

- Simultaneity is absolute—an observer-independent fact. And time must be continuous so that we can define velocity and acceleration. Classical time must also have an observer-independent notion of duration—what physicists call a metric—so that we can tell how far apart events are in time.

- Essentially, Newton proposed that the world comes equipped with a master clock that carves the world up into instants of time. Newton's physics listens to the ticking of this clock and no other. In classical physics, time flows and gives us an arrow pointing us toward the future.

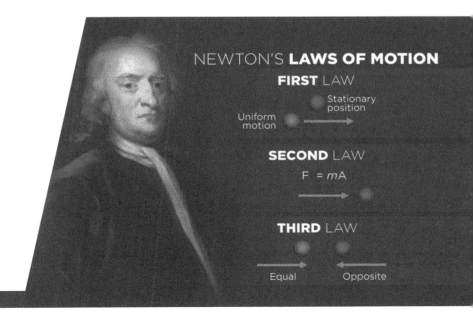

- Order, continuity, duration, simultaneity, flow, and the arrow are all logically detachable, but they all stick together in the master clock that Newton dubbed *time*. This model succeeded so well that it survived unscathed for almost two centuries.

CHALLENGES TO TIME

- Then came the challenges of the late 19th and early 20th centuries. The first was the work of Austrian physicist Ludwig Boltzmann, who reasoned that because Newton's laws work equally well going forward or backward in time, time has no built-in arrow. Instead, Boltzmann proposed that the distinction between past and future is not intrinsic to time but arises from asymmetries in how the matter in the universe is organized.

- Einstein mounted the next challenge by doing away with the idea of absolute simultaneity. According to his special theory of relativity, which events are happening at the same time depends on how fast an observer is moving.

- The true arena of events is neither time nor space but rather their union: spacetime. Two observers moving at different velocities disagree on when and where an event occurs, but they agree on its spacetime location.

- The year 1915 brought Einstein's general theory of relativity, which extends special relativity to situations where the force of gravity operates. Gravity distorts time such that a second's passage in one location may not mean the same thing as it does in another.

- Only in rare cases is it possible to synchronize clocks and have them stay synchronized, even in principle. In extreme situations, the world might not be divisible into instants of time at all. It then becomes impossible to say that an event happened before or after another.

- With that in mind, one might be tempted to think that the difference between space and time has nearly vanished and that the true arena of events in a relativistic universe is a big four-dimensional block. Relativity appears to spatialize time—that is, to turn it into merely one more direction within the block.

TIMELIKE AND SPACELIKE DIRECTIONS

- Even in general relativity, time retains a distinct and important function: that of distinguishing between timelike and spacelike directions. Timelike-related events are those that can be causally related. An object or signal can pass from one event to the other, influencing what happens. Spacelike-related events are causally unrelated.

- Observers disagree on the sequence of spacelike events, but they all agree on the order of timelike events. If one observer perceives that an event can cause another, all observers do.

- In his essay for a 2008 contest sponsored by the Foundational Questions Institute, philosopher Craig Callender explored what this feature of time means. Imagine slicing up spacetime from past to future. Each slice is the three-dimensional totality of space at one instant of time. The sum of all these slices of spacelike-related events is four-dimensional spacetime.

- Alternatively, imagine looking at the world sideways and slicing it up accordingly. From this perspective, each three-dimensional slice is a strange amalgam of events that are spacelike-related (in just two dimensions) and timelike-related.

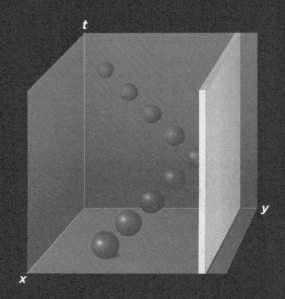

THE USUAL WAY takes slices of space at successive moments of time.

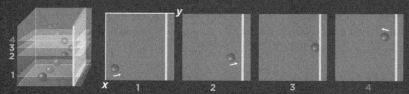

AN ALTERNATIVE considers slices not from past to future but from left to right.

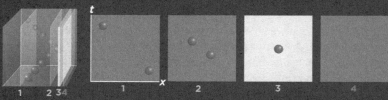

Lesson 4 | Is Time an Illusion?

- The first method is familiar to physicists and moviegoers alike. The frames of a movie represent slices of spacetime: They show space at successive moments of time. Like film buffs who instantly figure out the plot from just one scene, physicists can take a single complete spatial slice and reconstruct what happens on the other spatial slices, simply by applying the laws of physics.

- The second method of slicing has no simple analogy. Rather than slicing up spacetime from past to future, it involves carving from east to west. An example might be the north wall in a house plus what will happen on that wall in the future. From this slice, one could try to apply the laws of physics to reconstruct what the rest of the house (and indeed the rest of the universe) looks like.

- It's not immediately obvious whether the laws of physics allow such an analysis. But as mathematician Walter Craig of McMaster University and philosopher Steven Weinstein of the University of Waterloo have shown, it is possible, at least in some simple situations.

- In the normal, past-to-future slicing, the data needed from a slice are fairly easy to obtain. For instance, a physicist might measure the velocities of all particles. The velocity of a particle in one location is independent of the velocity of a particle someplace else, making both of them easy to measure.

- But in the second method, the particles' properties aren't independent; they have to be set up in a very specific way, or else a single slice wouldn't suffice to reconstruct all the others. A physicist would have to perform extremely difficult measurements on groups of particles to gather the data needed. Only in special cases would even these measurements allow reconstruction of the full spacetime.

- In a very precise sense, time is the direction within spacetime in which good prediction is possible—the direction in which we can tell the most informative stories. The narrative of the universe doesn't unfold in space. It unfolds in time.

THE CHALLENGE OF QUANTUM MECHANICS

- One of the highest goals of modern physics is to unite general relativity with quantum mechanics, producing a single theory that handles both the gravitational and quantum aspects of matter—a quantum theory of gravity. One of the stumbling blocks has been that quantum mechanics requires time to have properties that contradict information covered previously in this lesson.

- Quantum mechanics says that objects have a much richer repertoire of behaviors than we can possibly capture with classical quantities such as position and velocity. The full description of an object is given by a mathematical function called the quantum state, which evolves continuously in time. Using it, physicists are able to calculate the probabilities of any experimental outcome at any time.

- If we send an electron through a device that will deflect it either up or down, quantum mechanics may not be able to tell us with certainty which outcome to expect. Instead, the quantum state may give us only probabilities. An example would be giving a 25% chance the electron will veer upward and a 75% chance it'll veer downward. The outcomes of experiments are probabilistic. Two systems described with identical quantum states may give different outcomes.

- The theory's probabilistic predictions require time to have certain features. First, time is what makes contradictions possible. For instance, a rolled die can't have two different numbers facing up at the same time. Additionally, the probability of landing on each of the six numbers must add up to 100%.

- Second, the temporal order of quantum measurements makes a difference. Suppose a researcher passes an electron through a device that deflects it first along the vertical direction, then along the horizontal direction. As it emerges, the researcher measures its angular momentum. The researcher repeats the experiment, this time deflecting the electron horizontally, then vertically. The researcher then measures its angular momentum again. The values collected will be vastly different.

- Third, a quantum state provides probabilities for all of space at an instant of time. If the state encompasses a pair of particles, then measuring one particle instantaneously affects the other no matter where it is—leading to the infamous "spooky action at a distance" that so troubled Einstein. The reason it bothered him was that for the particles to react at the same time, the universe must have a master clock, which relativity expressly forbids.

UNIFICATION EFFORTS

- Physicists fret about the absence of time in relativity, but perhaps a worse problem is the central role of time in quantum mechanics. It's the reason unification has been so hard.

- A slew of research programs have sought to reconcile general relativity and quantum mechanics: superstring theory, causal triangulation theory, noncommutative geometry, and more. They split roughly into two groups.

- Physicists who think quantum mechanics provides the firmer foundation, like superstring theorists, start with a robust version of time. Those who believe general relativity provides the better starting point begin with a theory in which time is already demoted and hence are more open to the idea of a timeless reality.

▲ To convey the basic problem that time poses, focusing on the second approach is helpful. The leading instance of this strategy is loop quantum gravity, which descends from an earlier program known as canonical quantum gravity.

▲ Canonical quantum gravity emerged in the 1950s and 1960s, when physicists rewrote Einstein's equations for gravity in the same form as those for electromagnetism. The idea was that the techniques applied to electromagnetism could also be applied to gravity.

▲ When John Archibald Wheeler and Bryce DeWitt attempted this procedure in the late 1960s, they arrived at a very strange result. In the so-called Wheeler-DeWitt equation, shown below, the symbol t denoting time simply vanished.

$$\left[-G_{ijkl} \frac{\delta^2}{\delta\gamma_{ij}\delta\gamma_{kl}} - {}^3R(\gamma)\gamma^{1/2} + 2\Lambda\gamma^{1/2} \right] \Psi[\gamma_{ij}] = 0$$

$$G_{ijkl} = \frac{1}{2}\gamma^{-1/2}(\gamma_{ik}\gamma_{jl} + \gamma_{il}\gamma_{jk} - \gamma_{ij}\gamma_{kl})$$

▲ Decades of consternation followed for physicists: How could time just disappear? In retrospect, this result isn't very surprising, since time had already nearly disappeared from general relativity even before physicists attempted to merge it with quantum mechanics.

▲ If one takes this result literally, time doesn't really exist. Carlo Rovelli, one of the founders of loop quantum gravity, called his Foundational Questions Institute essay "Forget Time." He and English physicist Julian Barbour are the foremost proponents of this idea, and they've attempted to rewrite quantum mechanics in a timeless manner, as relativity appears to require.

▲ The reason they think this is possible is that general relativity still manages to describe change. It does so by relating physical systems directly to one another rather than to some abstract notion of global time.

Lesson 4 | Is Time an Illusion?

GETTING RID OF **TIME**

⊿ Although getting rid of time has its appeal, it inflicts collateral damage. For one, it requires quantum mechanics to be thoroughly rethought. Consider the famous case of Schrödinger's cat.

⊿ The cat is suspended between life and death, its fate hinging on the state of a quantum particle. In the usual way of thinking, the cat becomes alive or dead after a measurement takes place. But Rovelli would argue that the status of the cat is never resolved. The cat may be dead with respect to itself, alive relative to a human in the room, dead relative to a second human outside the room, and so on.

⊿ It's one thing to make the timing of the cat's death depend on the observer, as special relativity does. It's rather more surprising to make whether it even happens relative, as Rovelli suggests, following the spirit of relativity as far as it will go. Because time is so basic, banishing it would transform our worldview.

⊿ Even if the world is fundamentally timeless, it still seems to contain time. Anyone espousing timeless quantum gravity still has to explain why the world *seems* temporal. General relativity, too, lacks Newtonian time, but at least it has various partial substitutes that together behave like Newtonian time when gravity is weak and relative velocities low.

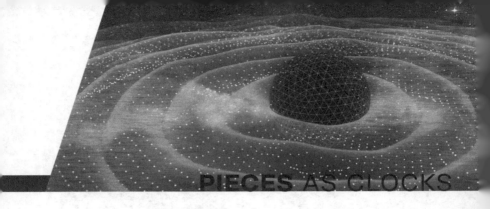

PIECES AS CLOCKS

- Canonical quantum gravity offers a more developed idea. Known as semiclassical time, it goes back to a 1931 paper by English physicist Nevill F. Mott describing the collision between a helium nucleus and a larger atom. To model the total system, Mott applied an equation lacking time that is usually applied only to static systems. He then divided the system into two subsystems and used the helium nucleus as a "clock" for the atom.

- Remarkably, relative to the nucleus, the atom obeys the standard time-dependent equation of quantum mechanics. A function of space plays the role of time. Even though the system as a whole is timeless, the individual pieces aren't. Hidden in the timeless equation for the total system is a time for the subsystem.

- Much the same works for quantum gravity, as Claus Kiefer of the University of Cologne argued in his Foundational Questions Institute essay. The universe may be timeless, but if one imagines breaking it into pieces, some of the pieces can serve as clocks for the others. Time emerges from timelessness. We perceive time because each of us is, by our very nature, one of those pieces.

- As interesting and startling as this idea is, it leaves us wanting more. The universe can't always be broken up into pieces that serve as clocks, and in those cases, the theory makes no probabilistic predictions. Handling such situations demands a full quantum theory of gravity and a deeper rethinking of time.

ABOUT THIS LESSON

This lesson was adapted from the articles "A Hole at the Heart of Physics" by George Musser, a science writer and contributing editor for *Scientific American*, and "Is Time an Illusion?" by Craig Callender, a philosophy professor at the University of California, San Diego.

Lesson 4 Transcript

IS TIME AN ILLUSION?

This lesson was adapted from the articles "A Hole at the Heart of Physics" by George Musser, a science writer and contributing editor for *Scientific American*, and "Is Time an Illusion?" by Craig Callender, a philosophy professor at the University of California, San Diego.

For most people, the great mystery of time is that there never seems to be enough of it. If it's any consolation, physicists have the same problem. The laws of physics contain a time variable called t, but it fails to capture key aspects of time as we live it, particularly the distinction between past and future. And as researchers try to formulate more fundamental laws, that little t evaporates altogether. Stymied, many physicists have sought help from an unfamiliar source: philosophers.

Philosophers? Yes, to most physicists, that sounds rather quaint. But philosophers have played a crucial role in past scientific revolutions, including the development of quantum mechanics and relativity in the early 20th century. Today a new revolution is under way, as physicists struggle to merge those two theories into a theory of quantum gravity that can reconcile two vastly different conceptions of space and time. Let's look at two examples of how physicists and philosophers have been pooling their resources.

The first concerns the problem of frozen time, also known simply as the problem of time. It arises when theorists try to turn Einstein's general theory of relativity into a quantum theory using what's called canonical quantization. The procedure worked brilliantly when applied to the theory of electromagnetism, but in the case of relativity, it produces the Wheeler-DeWitt equation, which lacks a time variable. Taken literally, the equation predicts that the universe should be frozen in time, but that's not what we see.

Some physicists and philosophers argue that this problem is connected to one of the founding principles of relativity, the idea that the laws of physics are the same for all observers. In the late 1980s, philosophers John Earman and John D. Norton, both at the University of Pittsburgh, argued that general relativity has startling implications for an old metaphysical question: Do space and time exist independently of stars, galaxies and everything else we perceive (this is a position known as substantivalism), or are space and time merely an artificial device to describe how physical objects are related (which is a principle known as relationism)? Or as Norton put it: "Are they like a canvas onto which an artist paints; they exist whether or not the artist paints on them? Or are they akin to parenthood; there is no parenthood until there are parents and children."

To find an answer, he and Earman revisited one of Einstein's thought experiments. Consider an empty patch of spacetime. Outside this hole, the distribution of matter fixes the geometry of spacetime, according to the equations of relativity, but inside, however, the equations let spacetime take on any of a variety of shapes. Spacetime behaves like a canvas tent. The tent poles, which represent matter, force the canvas to assume a certain shape. But if you leave out a pole, creating the equivalent of a hole, part of the tent can sag, or bow out, or ripple unpredictably in the wind and a similar ambiguity arises everywhere, not just in holes.

This thought experiment poses a dilemma. If the space-time continuum is a thing in its own right (as substantivalism holds), then general relativity must be indeterministic—that is, its description of the world must contain an element of randomness. For the theory to be deterministic, spacetime must be a mere fiction; and this is what relationism holds. So, at first glance, it looks like a victory for relationism, and it helps that other theories, such as electromagnetism, are based on symmetries that resemble relationism. But relationism has its own troubles.

It's actually the source of the problem of frozen time. Space may morph over time, but if its many shapes are all equivalent, it never truly changes. Plus, relationism clashes with the substantivalist underpinnings of quantum mechanics. If spacetime has no fixed meaning, how can you make observations at specific places and moments, as quantum mechanics seems to require? Different resolutions of the dilemma lead to very different theories of quantum gravity. Some are trying a relationist approach; they think time doesn't exist and have searched for ways to explain change as an illusion. Others, including string theorists, lean toward substantivalism.

"It's a good example of the value of philosophy for physics," says philosopher Craig Callender of the University of California, San Diego. "If physicists think the problem of time in canonical quantum gravity is solely a quantum problem, they're hurting their understanding of the problem—for it's been with us for much longer and is [a] more general [problem]."

So, now, a second example of philosophers' contributions concerns the arrow of time—the asymmetry of past and future. Many assume the arrow is explained by the second law of thermodynamics, which states that entropy, loosely defined as the amount of disorder within a system, increases with time. But nobody can really account for the second law.

The leading explanation, put forward by 19th-century Austrian physicist Ludwig Boltzmann, is probabilistic. The basic idea is that there are more ways for a system to be disordered than to be ordered. So, if the system is fairly ordered now, it'll probably be more disordered a moment from now. This reasoning, however, is symmetric in time. The system was probably more disordered a moment ago, too. As Boltzmann recognized, the only way to ensure that entropy will increase into the future is if it starts out with a low value in the past. So, the second law isn't so much a fundamental truth as historical happenstance, perhaps related to events early in the big bang.

Philosopher Huw Price of the University of Cambridge argues that almost every attempt to explain time asymmetry suffers some hidden presumption of time asymmetry. This circular reasoning also distorts how physicists interpret quantum mechanics. Price's work shows how philosophers can serve, in the words of philosopher Richard Healey of the University of Arizona, as the "intellectual conscience of the practicing physicist." Trained in logical rigor, they are experts at tracking down subtle biases. Life would be boring if we always listened to our conscience and physicists have often done best when ignoring philosophers. But in the eternal battle against our own leaps of logic, conscience is sometimes all we've got.

As you listen to me speak, you probably think this moment—right now—is what's happening. It feels special, it feels real. You remember the past and you anticipate the future, but you live in the present. Now that previous moment is no longer happening. This one is. To us, it feels like time flows. We have a deep intuition that the future is open until it becomes the present and that the past is fixed, and this structure gets carried forward as time flows on. It's built into our language, our thought and our behavior. How we live our lives depends on it.

Yet as natural as this way of thinking is, you won't find it reflected in science. Physics equations are like a map without a "you are here" symbol. The present moment doesn't exist in them and therefore neither does the flow of time. In fact, Einstein's theories of relativity suggest not only that there is no single special present but also that all moments are equally real. So, the future is no more open than the past.

The gap between our scientific understanding of time and our everyday experience has troubled thinkers throughout history and it has widened as physicists have gradually stripped time of most of the attributes we commonly ascribe to time. Now this rift is reaching its logical conclusion, for many theoretical physicists have come to believe that time fundamentally does not even exist.

The idea of a timeless reality is so startling at first that it's hard to see how it could be coherent. Everything we do, we do in time. The world is a series of events strung together by time. We see change and change is the variation of properties with respect to time. Without time, the world would be completely still. So, a timeless theory faces the challenge of explaining how we see change if the world isn't really changing.

Recent research attempts to pull off just this feat. Although time may not exist at a fundamental level, it may arise at higher levels—just as a table feels solid even though it's actually a swarm of particles composed mostly of empty space. Solidity is a collective, or an emergent, property of the particles. Time, too, could be an emergent property of whatever the basic ingredients of the world are. This concept of emergent time seems as revolutionary as the theories of relativity and quantum mechanics did a century ago. Einstein said that the key step forward in developing relativity was his reconceptualization of time. As physicists pursue his dream of unifying relativity with quantum mechanics, they believe time is again central.

In 2008, the Foundational Questions Institute sponsored an essay contest on the nature of time, and a veritable who's who of modern physicists weighed in. Many held that a unified theory will describe a timeless world. Others were loath to get rid of time. But the one thing they agreed on was that without thinking deeply about time, progress on unification may well be impossible. Our rich commonsensical notions of time have suffered a withering series of demotions throughout the ages. Time has many jobs to do in physics, but as physics has progressed, these jobs have been outsourced one by one.

It may not be obvious at first, but Newton's laws of motion require time to have many specific features. No matter when or where an event occurs, classical physics assumes all observers can objectively say whether it happens before, after or simultaneously with any other event in the universe. Simultaneity is absolute; it is an observer-independent fact. And time must be continuous so that we can define velocity and acceleration.

Classical time must also have a notion of duration—what physicists call a metric—so that we can tell how far apart events are in time. To say that Olympic sprinter Usain Bolt can run 27 miles per hour, we need a measure of what an hour is. Like the order of events, duration should be observer-independent. If Alice and Bob leave school at 3 pm, go their separate ways, and then meet back at home at 6 pm, the amount of time that has elapsed for Alice is equal to the amount of time that has elapsed for Bob.

So, basically, Newton proposed that the world comes equipped with a master clock that carves the world up into instants of time. Newton's physics listens to the ticking of this clock and no other. In classical physics, time flows and gives us an arrow pointing us toward the future. Order, continuity, duration, simultaneity, flow, the arrow—they're all logically detachable but they all stick together in the master clock that Newton dubbed *time*. This model succeeded so well that it survived unscathed for almost two centuries.

Then came the challenges of the late 19th and early 20th centuries. The first challenge was the work of Austrian physicist Ludwig Boltzmann, who reasoned that, because Newton's laws work equally well going forward and backward in time, time has no built-in arrow. Instead, Boltzmann proposed that the distinction between past and future is not intrinsic to time but arises from asymmetries in how the matter in the universe is organized. Although physicists still debate the details of this proposal, Boltzmann convincingly plucked away one feature of Newtonian time.

Einstein mounted the next challenge by doing away with the idea of absolute simultaneity. According to his special theory of relativity, which events are happening at the same time depends on how fast you're moving. The true arena of events is neither time nor space but rather their union: spacetime. So, two observers moving at different velocities disagree on when and where an event occurs, but they agree on its spacetime location. Space and time are secondary concepts that, as Einstein's university professor Hermann Minkowski declared, "Are doomed to fade away into mere shadows."

Things only get worse in 1915 with Einstein's general theory of relativity, which extends special relativity to situations where the force of gravity operates. Gravity distorts time, so that a second's passage here may not mean the same thing as it does there. Only in rare cases is it possible to synchronize clocks and have them stay synchronized, even in principle. We can't generally think of the world as unfolding, tick by tick, according to a single time parameter. In extreme situations, the world might not be divisible into instants of time at all. It then becomes impossible to say that an event happened before or after another.

So, what good is time, then? You might be tempted to think that the difference between space and time has nearly vanished and that the true arena of events in a relativistic universe is a big four-dimensional block. Relativity appears to spatialize time, to turn it into merely one more direction within the block.

Spacetime is like a loaf of bread that you can slice in different ways, called either *space* or *time* almost arbitrarily. Yet even in general relativity, time retains a distinct and important function: namely, that of distinguishing between *timelike* and *spacelike* directions. So, timelike-related events are those that can be causally related. An object or signal can pass from one event to the other, influencing what happens. Spacelike-related events are causally unrelated.

Mathematically, a mere minus sign differentiates the two directions, yet this minus sign has huge effects. Observers disagree on the sequence of spacelike events, but they all agree on the order of timelike events. If one observer perceives that an event can cause another, all observers should observe the same thing.

In his essay for the Foundational Questions Institute contest, philosopher Craig Callender (who is the person who wrote the article this segment is based on) explored what this feature of time really means. Imagine slicing up spacetime from past to future; each slice is the 3-D totality of space at one instant of time. The sum of all these slices of spacelike-related events is 4-D spacetime. Alternatively, imagine looking at the world sideways and slicing it up accordingly. From this perspective, each 3-D slice is a strange amalgam of events that are spacelike-related in two dimensions and timelike-related.

The first method is familiar to physicists and moviegoers. The frames of a movie represent slices of spacetime; they show space at successive moments of time. So, like a film buffs who instantly figure out the plot from just one scene, physicists can take a single complete spatial slice and reconstruct what happens on the other spatial slices, simply by applying the laws of physics.

But the second method of slicing has no simple analogy. Rather than slicing up spacetime from past to future, we're carving from east to west. An example might be the north wall in your house plus what will happen on that wall in the future. From this slice, you could try to apply the laws of physics to reconstruct what the rest of your house (and indeed the rest of the universe) looks like.

If that sounds strange, it should. It's not immediately obvious whether the laws of physics let you do this. But as mathematician Walter Craig of McMaster University and philosopher Steven Weinstein of the University of Waterloo have shown, you can, at least in some simple situations.

Although both methods of slicing are possible in principle, they're profoundly different. In the normal, past-to-future slicing, the data you need to collect on a slice are fairly easy to obtain. For instance, you measure the velocities of all particles. The velocity of a particle in one location is independent of the velocity of a particle someplace else, making both of them easy to measure.

But in the second method, the particles' properties aren't independent; they have to be set up in a very specific way, or else a single slice wouldn't suffice to reconstruct all the others. You'd have to perform extremely difficult measurements on groups of particles to gather the data you need and only in special cases would you even be allowed to use these measurements to reconstruct the full spacetime.

In a very precise sense, time is the direction within spacetime in which good prediction is possible, the direction in which we can tell the most informative stories. The narrative of the universe doesn't unfold in space. It unfolds in time.

One of the highest goals of modern physics is to unite general relativity with quantum mechanics, producing a single theory that handles both the gravitational and quantum aspects of matter—a quantum theory of gravity. One of the stumbling blocks has been that quantum mechanics requires time to have properties that contradict what we've discussed so far.

Quantum mechanics says that objects have a much richer repertoire of behaviors than we can possibly capture with classical quantities such as position and velocity. The full description of an object is given by a mathematical function called the quantum state, which evolves continuously in time. Using it, physicists are able to calculate the probabilities of any experimental outcome at any time.

So, for instance, if we send an electron through a device that will deflect it either up or down, quantum mechanics may not be able to tell us with certainty which outcome to expect. Instead the quantum state may give us only probabilities; say, a 25% chance the electron will veer upward and a 75% chance it'll veer downward. The outcomes of experiments are probabilistic. So, two systems described with identical quantum states may give very different outcomes.

The theory's probabilistic predictions require time to have certain features. First, time is what makes contradictions possible. A rolled die can't have both a five and a three facing up at the same time. Also, the probability of landing on each of the six numbers must add up to 100%; otherwise the concept of probability wouldn't be meaningful. The probabilities add up at a time, not at a place. The same is true of the probabilities for quantum particles to have a given position or momentum.

Second, the temporal order of quantum measurements makes a difference. Suppose I pass an electron through a device that deflects it first along the vertical direction, then along the horizontal direction. As it emerges, I measure its angular momentum. Then I repeat the experiment, this time deflecting the electron horizontally, then vertically, and measure its angular momentum again. The values I get will be vastly different.

Third, a quantum state provides probabilities for all of space at an instant of time. If the state encompasses a pair of particles, then measuring one particle instantaneously affects the other no matter where it is; and this leads to the infamous "spooky action at a distance" that really troubled Einstein. The reason it bothered him was that for the particles to react at the same time, the universe must have a master clock, which relativity expressly forbids.

Although some of these issues are controversial, time in quantum mechanics is basically a throwback to Newtonian mechanics. Physicists fret about the absence of time in relativity, but perhaps a worse problem is the central role of time in quantum mechanics. It is the reason unification has been so hard.

A slew of research programs have sought to reconcile general relativity and quantum mechanics, including superstring theory, causal triangulation theory, noncommutative geometry, and more. They split roughly into two groups. Physicists who think quantum mechanics provides the firmer foundation—these include superstring theorists—and they start with a robust version of time. But those who believe general relativity provides the better starting point begin with a theory in which time is already demoted and hence are more open to the idea of a timeless reality.

To convey the basic problem that time poses, let's focus on the second approach. The leading instance of this strategy is loop quantum gravity, which descends from an earlier program known as canonical quantum gravity. Canonical quantum gravity emerged in the 1950s and 60s, when physicists rewrote Einstein's equations for gravity in the same form as those for electromagnetism. The idea was that the techniques applied to electromagnetism could also be applied to gravity.

When John Archibald Wheeler and Bryce DeWitt attempted this procedure in the late 1960s, they arrived at a very strange result. In the so-called Wheeler-DeWitt equation, the symbol t denoting time simply vanished. Decades of consternation followed for physicists. How could time just disappear? In retrospect, this result isn't very surprising, since time had already nearly disappeared from general relativity even before physicists attempted to merge it with quantum mechanics.

If you take this result literally, time doesn't really exist. Carlo Rovelli, one of the founders of loop quantum gravity, called his Foundational Questions Institute essay "Forget Time." He and English physicist Julian Barbour are the foremost proponents of this idea, and they've attempted to rewrite quantum mechanics in a timeless manner, as relativity appears to require.

The reason they think this is possible is that general relativity still manages to describe change. It does so by relating physical systems directly to one another rather than to some abstract notion of global time. Instead of describing an event with time, we can correlate it with a satellite's orbit. Instead of saying a baseball accelerates at 10 meters per second per second, we can describe it in terms of the change of a glacier. And so on. Time becomes redundant. Change can be described without it.

This vast network of correlations is neatly organized so that we can define something called *time* and relate everything to it, relieving ourselves of the burden of keeping track of all those direct relations. But this convenient fact shouldn't trick us into thinking that time is a fundamental part of the world's furniture.

So, similarly, money makes life much easier than bartering every time you just want a coffee, but it's just an invented placeholder for the things we value, not something we value in and of itself. Similarly, time allows us to relate physical systems to one another without having to figure out exactly how a glacier relates to a baseball. But it, too, may be a convenient fiction that no more exists fundamentally in the natural world than money does.

Although getting rid of time has its appeal, it inflicts a good deal of collateral damage. For one, it requires quantum mechanics to be thoroughly rethought. Consider the famous case of Schrödinger's cat. The cat is suspended between life and death, its fate hinging on the state of a quantum particle. In the usual way of thinking, the cat becomes alive or dead after a measurement takes place. But Rovelli would argue that the status of the cat is never resolved. The poor thing may be dead with respect to itself, alive relative to a human in the room, dead relative to a second human outside the room, and so on.

It's one thing to make the timing of the cat's death depend on the observer, as special relativity does. It's rather more surprising to make whether it even happens relative, as Rovelli suggests, following the spirit of relativity as far as it will go. Because time is so basic, banishing it would transform our worldview.

Even if the world is fundamentally timeless, it still seems to contain time. Anyone espousing timeless quantum gravity still has to explain why the world seems temporal. General relativity, too, lacks Newtonian time, but at least it has various partial substitutes that together behave like Newtonian time when gravity is weak and relative velocities are low.

Canonical quantum gravity offers a more developed idea. Known as semiclassical time, it goes back to a 1931 paper by English physicist Nevill

F. Mott describing the collision between a helium nucleus and a larger atom. To model the total system, Mott applied an equation lacking time that is usually applied only to static systems. He then divided the system into two subsystems and used the helium nucleus as a clock for the atom.

Remarkably, relative to the nucleus, the atom obeys the standard time-dependent equation of quantum mechanics. A function of space plays the role of time. So, even though the system as a whole is timeless, the individual pieces aren't. So hidden in the timeless equation for the total system is a time for the subsystem.

Much the same works for quantum gravity, as Claus Kiefer of the University of Cologne argued in his Foundational Questions Institute essay. The universe may be timeless, but if you imagine breaking it into pieces, some of the pieces can serve as clocks for the others. Time emerges from timelessness. We perceive time because each of us is, by our very nature, one of those pieces.

So as interesting and startling as this idea is, it leaves us wanting more. The universe can't always be broken up into pieces that serve as clocks, and in those cases, the theory makes no probabilistic predictions. Handling such situations demands a full quantum theory of gravity and a deeper rethinking of time.

Historically, physicists began with the highly structured time of experience, with a fixed past, present and an open future. They gradually dismantled this structure, and little, if any, of it remains. Researchers must now reverse this train of thought and reconstruct the time of experience from the time of nonfundamental physics, which itself may need to be reconstructed from a network of correlations among pieces of a fundamental static world.

French philosopher Maurice Merleau-Ponty argued that time's apparent flow is just a product of our "surreptitiously putting into the river a witness of its course." That is, the tendency to believe time flows is a result of forgetting to put ourselves and our connections to the world into the picture.

Merleau-Ponty was speaking of our subjective experience of time, and until recently no one ever guessed that objective time might itself be explained as a result of those connections. Time may exist only by breaking the world into subsystems and looking at what ties them together. In this picture, physical time emerges by virtue of our thinking of ourselves as separate from everything else.

5

TIME TRAVEL AND THE TWIN PARADOX

Lesson 5 | Time Travel and the Twin Paradox

To look at the topic of time travel, this lesson explores the roles of speed and gravity. Additionally, the lesson speculates on wormholes as a tool for time travel as well as paradoxes presented by time travel, including the famous twin paradox.

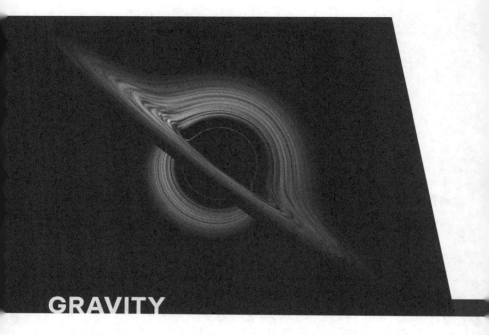

GRAVITY

- In his general theory of relativity, Einstein predicted that gravity slows time. This means that clocks run faster in space than they do on the ground. The effect is minuscule, but it has been directly measured using accurate clocks. In fact, these time-warping effects have to be taken into account in the Global Positioning System.

- At the surface of a neutron star, gravity is so strong that time is slowed down by about 30% relative to Earth time. Viewed from such a star, events on Earth would look like a fast-forwarded video.

- A black hole represents the ultimate time warp. At the surface of the hole, time stands still relative to Earth. This means that if an astronaut fell into a black hole from nearby, in the brief interval it took her to reach the surface, all of eternity would pass by in the wider universe. However, if an astronaut could zoom very close to a black hole and return unscathed—admittedly a fanciful and foolhardy prospect—she could leap far into the future.

Lesson 5 | Time Travel and the Twin Paradox

GOING BACKWARD

- Going backward in time is much more problematic. In 1948, Kurt Gödel of the Institute for Advanced Study produced a solution of Einstein's gravitational field equations that described a rotating universe. In this universe, an astronaut could travel through space so as to reach his own past.

- This is because of the way gravity affects light. The rotation of the universe would drag light around with it, allowing a material object to travel in a closed loop in space that's also a closed loop in time, without at any point exceeding the speed of light in the immediate neighborhood of the particle.

- Gödel's solution was shrugged off as a mathematical curiosity. Still, his result demonstrated that going back in time isn't forbidden by the theory of relativity.

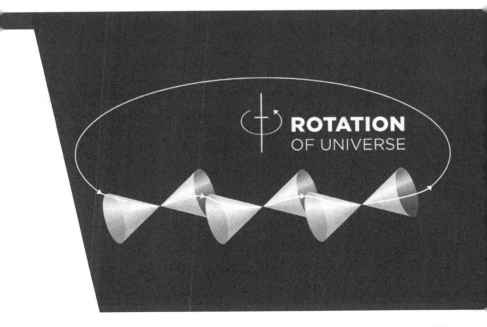

- Other scenarios have been found to permit travel into the past as well. For example, in 1974, Frank J. Tipler of Tulane University calculated that a massive, infinitely long cylinder spinning on its axis at near the speed of light could let astronauts visit their own past, again by dragging light around the cylinder into a loop.

- In the mid-1980s, the most realistic scenario for a time machine emerged. This one was based on the concept of a wormhole. Because they offer a shortcut between two widely separated points in space, in science fiction, wormholes are sometimes called stargates. An astronaut going through a hypothetical wormhole might come out moments later on the other side of the galaxy.

Lesson 5 | Time Travel and the Twin Paradox

- Wormholes fit into the general theory of relativity, which says gravity warps not only time but also space. The theory allows the analogue of alternative road and tunnel routes connecting two points in space. Mathematicians refer to such a space as multiply connected. Just as a tunnel passing through a hill can be shorter than the street along the surface of the hill, a wormhole may be shorter than the usual route through ordinary space.

- After Carl Sagan used the wormhole as a fictional device in his novel *Contact*, Kip Thorne and his colleagues at the California Institute of Technology set out to discover whether wormholes were consistent with known physics. Their starting point was the idea that a wormhole, like a black hole, would be an object with fearsome gravity. Yet unlike a black hole, which offers a one-way journey to nowhere, a wormhole would have an exit as well as an entrance.

- For the wormhole to be traversable, though, it must contain what Thorne called "exotic matter." In effect, this is something that will generate antigravity to combat the natural tendency of a massive system to implode into a black hole under its own intense weight.

- Antigravity, or gravitational repulsion, can be generated by negative energy or pressure. Negative-energy states are known to exist in certain quantum systems, which suggests that Thorne's exotic matter isn't ruled out by the laws of physics. However, it's unclear whether enough antigravitating stuff can be assembled to stabilize a wormhole.

- Soon Thorne and his colleagues realized that if a stable wormhole could be created, then it could readily be turned into a time machine. An astronaut who passed through one might come out not only somewhere else in the universe but at another time, too—in either the future or the past.

- A formidable problem that stands in the way of making a wormhole time machine is the creation of the wormhole in the first place. Possibly space is threaded with such structures naturally—relics of the big bang. If so, a supercivilization might commandeer one.

- Alternatively, wormholes might naturally come into existence on tiny scales, the so-called Planck length. In principle, such a minute wormhole could be stabilized by a pulse of energy and then somehow inflated to usable dimensions.

TIME TRAVEL STRANGENESS

- Assuming the engineering problems could be overcome, the production of a time machine would open up a Pandora's box of paradoxes. Consider, for example, this one: A billiard ball passes through a wormhole time machine. When it emerges, it hits its earlier self, thereby preventing it from ever entering the wormhole.

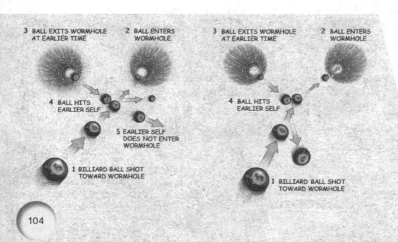

- Resolution of the paradox proceeds from a simple realization: The billiard ball can't do anything that's inconsistent with logic or with the laws of physics. It can't pass through the wormhole in such a way that will prevent it from passing through the wormhole. Yet nothing stops it from passing through the wormhole in an infinity of other ways.

- Even if time travel isn't strictly paradoxical, it certainly is strange. Consider a time traveler who leaps ahead a year and reads about a new mathematical theorem in a future edition of *Scientific American*. She notes the details, returns to her own time, and teaches the theorem to a student, who then writes it up for *Scientific American*. The article is, of course, the very one that the time traveler read.

- The question then arises: Where did the information about the theorem come from? Not from the time traveler, because she read it. But it did not come from the student either, who learned it from the time traveler. The information seems to come into existence from nowhere, without reason.

- The bizarre consequences of time travel have led some scientists to reject the notion altogether. Stephen Hawking of the University of Cambridge proposed a "chronology protection conjecture," which would outlaw causal loops.

- Because the theory of relativity is known to permit causal loops, chronology protection would require some other factor to intercede to prevent travel into the past. What might this factor be? One suggestion is that quantum processes will come to the rescue.

- The existence of a time machine would allow particles to loop into their own past. Calculations hint that the ensuing disturbance would become self-reinforcing, creating a runaway surge of energy that would wreck the wormhole. Chronology protection is still just a conjecture, so time travel remains a possibility.

TIME AND THE TWIN PARADOX

- In his special theory of relativity, Albert Einstein proposed that the measured interval between two events depends on how the observer is moving. Crucially, two observers who move differently will experience different durations between the same two events.

- The term *time dilation* was coined to describe the slowing of time caused by motion. To illustrate the effect of time dilation, Einstein proposed an example—the twin paradox—that is arguably the most famous thought experiment in relativity theory.

- To illustrate the paradox: One of two twins travels at near the speed of light to a distant star and returns to Earth. Relativity dictates that when the sister comes back, she's younger than her brother. The paradox lies in the question: Why is the traveling twin younger?

- Special relativity tells us that an observed clock, traveling at high speed past an observer, appears to run more slowly—that is, it experiences time dilation. Since special relativity says there's no absolute motion, wouldn't the sister traveling to the star also see her brother's clock on Earth move more slowly? If this were the case, wouldn't they both be the same age?

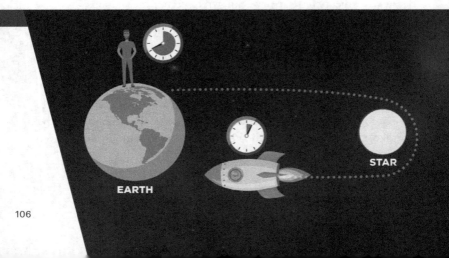

Lesson 5 | Time Travel and the Twin Paradox

- This paradox is typically explained by saying that the twin who feels the acceleration is the one who's younger at the end of the trip; hence, the sister who travels to the star is younger. Although the result is correct, the explanation is misleading.

- Some people may incorrectly assume that the acceleration causes the age difference. But the acceleration incurred by the traveler is incidental, and the paradox can be unraveled by special relativity.

UNRAVELING THE PARADOX

- Assume that the twins, nicknamed the traveler and the homebody, live in Hanover, New Hampshire. They share a common desire to build a spacecraft that can achieve 0.6 times the speed of light (0.6 c). After working on the spacecraft for years, they're ready to launch it, piloted by the traveler, toward a star six light-years away. Her vehicle will quickly accelerate to 0.6 c.

- The traveler uses the length-contraction equation of special relativity to measure distance. The star six light-years away to the homebody appears to be only 4.8 light-years away to the traveler at a speed of 0.6 c. Thus, to the traveler, the trip to the star takes only eight years (derived by dividing 4.8 by 0.6), whereas the homebody calculates it taking 10 years (derived by dividing 6.0 by 0.6).

- Assume each twin has a very powerful telescope that permits observation of the other. Both the traveler and the homebody set their clocks at zero when the traveler leaves Earth for the star. When the traveler reaches the star, her clock reads eight years. But when the homebody sees the traveler reach the star, the homebody's clock reads 16 years.

▲ The reason is this: To the homebody, the craft takes 10 years to make it to the star, and the light takes six additional years to come back to Earth, showing the traveler at the star. Viewed through the homebody's telescope, the traveler's clock appears to be running at half the speed of his clock.

▲ As the traveler reaches the star, she reads her clock at eight years as mentioned, but she sees the homebody's clock as it was six years ago (the amount of time it takes for the light from Earth to reach her). She sees it as four years (10 minus 6). The traveler also views the homebody's clock as running at half the speed of her clock (derived by dividing 4 by 8).

▲ On the trip back, the homebody views the traveler's clock going from eight years to 16 years in only four years' time because his clock was at 16 years when he saw the traveler leave the star, and it will be at 20 years when the traveler arrives back home. The homebody sees the traveler's clock advance eight years in four years of his time; it's now running twice as fast as his clock.

▲ As the traveler returns home, she sees the homebody's clock advance from four to 20 years in eight years of his time. She also sees her brother's clock advancing at twice the speed of her own. They both agree, however, that at the end of the trip the traveler's clock reads 16 years and the homebody's 20 years. Therefore, the traveler is four years younger.

▲ The asymmetry in the paradox is that the traveler leaves Earth's reference frame and comes back, whereas the homebody never leaves Earth. It's also an asymmetry that the traveler and the homebody agree with the reading on the traveler's clock at each event but that they don't agree about the reading on the homebody's clock at each event. The traveler's actions define the events.

▲ The Doppler effect and relativity together explain this effect mathematically at any instant. Note, too, that the speed at which an observed clock appears to run depends on whether it's traveling away from or toward the observer.

Lesson 5 | Time Travel and the Twin Paradox

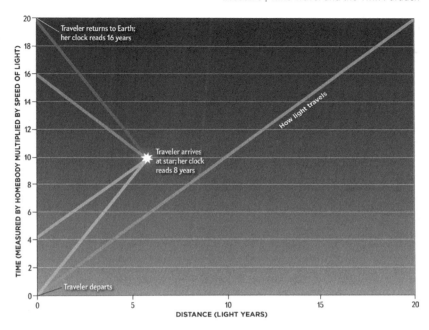

▲ The fundamentals of this explanation have been exhaustively confirmed experimentally. The twin paradox today is more than just a theory.

ABOUT THIS LESSON

This lesson was adapted from the articles "How to Build a Time Machine" by Paul Davies, a theoretical physicist and cosmologist at Arizona State University, and "Time and the Twin Paradox" by Ronald C. Lasky, an instructional professor at Dartmouth College and a senior technologist at Indium Corporation.

Lesson 5 Transcript

TIME TRAVEL AND THE TWIN PARADOX

This lesson was adapted from the articles "How to Build a Time Machine" by Paul Davies, a theoretical physicist and cosmologist at Arizona State University, and "Time and the Twin Paradox" by Ronald C. Lasky, an instructional professor at Dartmouth College and a senior technologist at Indium Corporation.

Time travel has been a popular science-fiction theme since H. G. Wells wrote his celebrated novel *The Time Machine* in 1895. But is it really possible to build a machine that can transport a human being into the past or the future?

For decades time travel stood on the fringe of respectable science. But in recent years, the topic has become something of a cottage industry among theoretical physicists. The motivation has been partly recreational—time travel is fun to think about. But this research has a serious side, too. Understanding the relation between cause and effect is a key part of attempts to construct a unified theory of physics. If unrestricted time travel were possible, even in principle, it would have profound consequences for our understanding of the nature of the world.

Our best understanding of time comes from Albert Einstein's theories of relativity. Before Einstein, time was widely understood to be absolute and universal, the same for everyone no matter their physical circumstances. But in his special theory of relativity, Einstein proposed that the measured interval between two events depends on how the observer is moving. Crucially, two observers who move differently will experience different durations between the same two events. The effect is often described using what's called the twin paradox.

Suppose Sally and Sam are twins. Sally boards a rocket ship and travels at high speed to a nearby star, turns around and flies back to Earth, while Sam stays at home. For Sally, the duration of the journey might be, say, one year, but when she returns and steps out of the spaceship, she finds that 10 years have elapsed on Earth. Her brother is now nine years older than she is. Despite the fact that they were born on the same day, Sally and Sam are no longer the same age. This example illustrates a limited type of time travel. In effect, Sally has leaped nine years into Earth's future.

This effect, known as time dilation, occurs whenever two observers move relative to each other. In daily life, we don't notice weird time warps, because the effect becomes dramatic only when the motion occurs at close to the speed of light. Even at aircraft speeds, the time dilation in a typical journey amounts to just a few nanoseconds; so, hardly an adventure of Wellsian proportions. Still, atomic clocks are accurate enough to record the shift and confirm that time really is stretched by motion. So travel into the future is a proven fact, even if it has so far been in rather unexciting amounts.

To observe really dramatic time warps, we need to look beyond the realm of ordinary experience. Subatomic particles can be propelled at nearly the speed of light in large accelerator machines. Some of these particles, such as muons, have a built-in clock because they decay with a definite half-life. In accordance with Einstein's theory, fast-moving muons inside accelerators are observed to decay in slow motion.

Some cosmic rays also experience spectacular time warps. These particles move so close to the speed of light that, from their point of view, they cross the galaxy in minutes, even though in Earth's frame of reference they seem to take tens of thousands of years. If time dilation didn't occur, these particles would never make it here. So, speed's one way to jump ahead in time and gravity is another.

In his general theory of relativity, Einstein predicted that gravity slows time. This means that clocks run a bit faster in the attic than they do in the basement, because the basement is closer to the center of Earth and therefore deeper down in a gravitational field. Similarly, clocks run faster in space than they do on the ground. Once again, the effect is minuscule, but it's been directly measured using accurate clocks. In fact, these time-warping effects have to be taken into account in the Global Positioning System. If they aren't, boaters and drivers could find themselves many kilometers off course.

So, imagine the surface of a neutron star where gravity is so strong that time is slowed down by about 30% relative to Earth time. Viewed from such a star, events here would look like a fast-forwarded video. Now, a black hole represents the ultimate time warp. At the surface of the hole, time stands still relative to Earth. This means that if you fell into a black hole from nearby, in the brief interval it took you to reach the surface, all of eternity would pass by in the wider universe.

Therefore, as far as the outside universe is concerned, the region within the black hole is beyond the end of time. If an astronaut could zoom very close to a black hole and return unscathed (which is admittedly a fanciful and foolhardy, prospect) she could leap far into the future.

Okay, so far, we've talked about traveling forward in time. But what about going backward? That is much more problematic. In 1948, Kurt Gödel of the Institute for Advanced Study in Princeton, New Jersey, produced a solution of Einstein's gravitational field equations that described a rotating universe. In his universe, an astronaut could travel through space so as to reach his own past. This is because of the way gravity affects light. The rotation of the universe would drag light around with it, allowing a material object to travel in a closed loop in space that's also a closed loop in time, without at any point exceeding the speed of light in the immediate neighborhood of the particle.

Gödel's solution was shrugged off as a mathematical curiosity; after all, observations show no sign that the universe as a whole is spinning. Still, his result demonstrated that going back in time is not forbidden by the theory of relativity. Indeed, Einstein confessed he was troubled by the thought that his theory might permit travel into the past under some circumstances.

Other scenarios have been found to permit travel into the past, as well. For example, in 1974, Frank J. Tipler of Tulane University calculated that a massive, infinitely long cylinder spinning on its axis at near the speed of light could let astronauts visit their own past, again by dragging light around the cylinder into a loop.

In the mid-1980s, the most realistic scenario for a time machine emerged and this one was based on the concept of a wormhole. Because they offer a shortcut between two widely separated points in space, in science fiction wormholes are sometimes called stargates. Jump through a hypothetical wormhole and you might come out moments later on the other side of the galaxy. Wormholes fit into the general theory of relativity, which says gravity warps not only time but also space. Just as a tunnel passing through a hill can be shorter than the street along the surface of the hill, a wormhole may be shorter than the usual route through ordinary space.

After Carl Sagan used the wormhole as a fictional device in his novel *Contact*, Kip Thorne and his colleagues at the California Institute of Technology set out to discover whether wormholes were consistent with known physics. Their starting point was the idea that a wormhole, like a black hole, would be an object with fearsome gravity. Yet unlike a black hole, which offers a one-way journey to nowhere, a wormhole would have an exit as well as an entrance.

For the wormhole to be traversable, though, it must contain what Thorne called "exotic matter." In effect, this is something that will generate antigravity to combat the natural tendency of a massive system to implode into a black hole under its own intense weight. Antigravity, or gravitational repulsion, can be generated by negative energy or pressure. Negative-energy states are known to exist in certain quantum systems, which suggests that Thorne's exotic matter isn't ruled out by the laws of physics. That said, it's unclear whether enough antigravitating stuff can be assembled to stabilize a wormhole.

Soon Thorne and his colleagues realized that if a stable wormhole could be created, then it could readily be turned into a time machine. An astronaut who passed through one might come out not only somewhere else in the universe but somewhen else, too— either the future or in the past.

To adapt the wormhole for time travel, one of its mouths could be towed to a neutron star and placed close to its surface. The gravity of the star would slow time near that wormhole mouth, so a time difference between the ends of the wormhole would gradually accumulate. If both mouths were then parked at a convenient place in space, this time difference would remain frozen in.

So, suppose the difference were 10 years. An astronaut passing through the wormhole in one direction would jump 10 years into the future, whereas an astronaut passing in the other direction would jump 10 years into the past. By returning to her starting point at high speed across ordinary space, the second astronaut might get back home before she left. In other words, a closed loop in space could become a loop in time as well. The one restriction is that the astronaut could not return to a time before the wormhole was first built.

A formidable problem that stands in the way of making a wormhole time machine is the creation of the wormhole in the first place. Possibly space is threaded with such structures naturally; perhaps they are relics of the big bang and if so, a supercivilization might commandeer one.

Alternatively, wormholes might naturally come into existence on tiny scales, the so-called Planck length, which is about 20 factors of 10 as small as an atomic nucleus. In principle, such a minute wormhole could be stabilized by a pulse of energy and then somehow inflated to usable dimensions.

So, let's tie it all together and see how we can build a wormhole time machine in three steps. Step 1: Find or build a wormhole. Large wormholes might exist naturally in deep space, a relic of the big bang. Otherwise, we'd have to make do with subatomic wormholes—either natural ones (which are thought to be winking in and out of existence all around us) or from artificial ones (which are produced by particle accelerators). These smaller wormholes would have to be enlarged to useful size, perhaps using energy fields like those that caused space to inflate shortly after the big bang.

Step 2: Stabilize the wormhole. An infusion of negative energy, produced by quantum means such as the so-called Casimir effect, would allow a signal or object to pass safely through the wormhole. Negative energy counteracts the tendency of a wormhole to pinch off into a point of infinite or near-infinite density and become a black hole.

Step 3: Tow the wormhole. A spaceship, presumably of highly advanced technology, would separate the mouths of the wormhole. One mouth might be positioned near the surface of a neutron star, an extremely dense star with a strong gravitational field. The intense gravity causes time to pass more slowly. Because time passes more quickly at the other wormhole mouth, the two mouths become separated not only in space but also in time.

Voilà. A wormhole time machine in just three steps! Easy, right? Okay, yeah, maybe not.

Now, assuming the engineering problems could be overcome, the production of a time machine would open up a Pandora's box of paradoxes. Consider, for example, the time traveler who visits the past and murders his own mother when she's still a young girl. How do we make sense of this? If the girl dies, she can't become the time traveler's mother, and if the time traveler was never born, he couldn't go back and murder his mother.

A simplified version of this paradox involves billiard balls. A billiard ball passes through a wormhole time machine. Upon emerging, it hits its earlier self, thereby preventing it from ever entering the wormhole. Resolution of the paradox proceeds from a simple realization: The billiard ball can't do anything that's inconsistent with logic or with the laws of physics; so, it can't pass through the wormhole in such a way that will prevent it from passing through the wormhole. Yet nothing stops it from passing through the wormhole in an infinity of other ways.

Even if time travel isn't strictly paradoxical, it certainly is weird. Consider the time traveler who leaps ahead a year and reads about a new mathematical theorem in a future edition of *Scientific American*. She notes the details, returns to her own time and teaches the theorem to a student, who then writes it up for *Scientific American*. The article is, of course, the very one that the time traveler read. So, the question then arises: Where did the information about the theorem come from? Not from the time traveler, because she read it, but also not from the student, who learned it from the time traveler. The information seems to come into existence from nowhere, without reason.

The bizarre consequences of time travel have led some scientists to reject the notion altogether. Stephen Hawking of the University of Cambridge proposed a "chronology protection conjecture," which would outlaw causal loops. Because the theory of relativity is known to permit causal loops, chronology protection would require some other factor to intercede to prevent travel into the past.

So, what might this factor be? One suggestion is that quantum processes will come to the rescue. The existence of a time machine would allow particles to loop into their own past. Calculations hint that the ensuing disturbance would become self-reinforcing, creating a runaway surge of energy that would wreck the wormhole.

Chronology protection is still just a conjecture, so time travel remains a possibility. It is even conceivable that the next generation of particle accelerators—or perhaps the existing Large Hadron Collider at CERN near Geneva—will be able to produce subatomic collisions powerful enough to rip spacetime itself, creating wormholes that survive long enough for nearby particles to execute fleeting causal loops. This would be a far cry from Wells's vision of a time machine, but it would forever change our picture of physical reality.

As maxims go, "time is relative" may not be quite as famous as "time is money," but the notion that time speeds up or slows down depending on how fast one object is traveling relative to another surely ranks as one of Albert Einstein's most inspired insights. The term "time dilation" was coined to describe the slowing of time caused by motion. To illustrate the effect of time dilation, Einstein proposed an example—called the twin paradox—that is arguably the most famous thought experiment in relativity theory.

As a reminder, one of two twins travels at near the speed of light to a distant star and returns to Earth. Relativity dictates that when she comes back, she's younger than her brother. The paradox lies in the question, "Why is the traveling twin younger?" So, special relativity tells us that an observed clock, traveling at high speed past an observer, appears to run more slowly—that is, it experiences time dilation.

Since special relativity says there's no absolute motion, wouldn't the sister traveling to the star also see her brother's clock on Earth move more slowly? If this were the case, wouldn't they both be the same age? This paradox is typically explained by saying that the twin who feels the acceleration is the one who's younger at the end of the trip; hence, the sister who travels to the star is younger. Although the result is correct, the explanation is misleading. Some people may incorrectly assume that the acceleration causes the age difference. But the acceleration incurred by the traveler is incidental and the paradox can be unraveled by special relativity.

Let's assume that the twins, nicknamed the traveler and the homebody, live in Hanover, New Hampshire. They differ in their wanderlust but share a common desire to build a spacecraft that can achieve 0.6 times the speed of light (which we call 0.6 c). After working on the spacecraft for years, they're ready to launch it, piloted by the traveler, toward a star six light-years away.

Her vehicle will quickly accelerate to 0.6 c. To reach that speed, it would take a little more than say a 100 days at an acceleration of two g's. Two g's is two times the acceleration of gravity, about what one experiences on a sharp loop of a roller coaster. If the traveler were an electron, she could be accelerated to 0.6 c in a tiny fraction of a second. So, the time to reach 0.6 c isn't really central to the argument.

The traveler uses the length-contraction equation of special relativity to measure distance. So the star six light-years away to the homebody appears to be only 4.8 light-years away to the traveler at a speed of 0.6 c. Thus, to the traveler, the trip to the star takes only eight years (4.8 divided by 0.6), whereas the homebody calculates it taking 10 years (or 6.0 divided by 0.6).

To solve the twin paradox, we need to consider how each twin would view their own and the other's clocks during the trip. So, let's assume each twin has a very powerful telescope that permits such observation. Surprisingly, by focusing on the time it takes light to travel between the two, the paradox can be explained.

Both the traveler and the homebody set their clocks at zero when the traveler leaves Earth for the star. When the traveler reaches the star, her clock reads 8 years. But when the homebody sees the traveler reach the star, the homebody's clock reads 16 years. Why 16 years? Because, to the homebody, the craft takes 10 years to make it to the star, and the light takes 6 additional years to come back to Earth, showing the traveler at the star. So, viewed through the homebody's telescope, the traveler's clock appears to be running at half the speed of his clock (or 8 divided by 16).

As the traveler reaches the star, she reads her clock at 8 years as mentioned, but she sees the homebody's clock as it was 6 years ago (the amount of time it takes for the light from Earth to reach her); so, that is, at 4 years (10 minus 6). So, the traveler also views the homebody's clock as running at half the speed of her clock (4 divided by 8).

On the trip back, the homebody views the traveler's clock going from 8 years to 16 years in only 4 years' time because his clock was at 16 years when he saw the traveler leave the star, and it will be at 20 years when the traveler arrives back home. So, the homebody sees the traveler's clock advance 8 years in 4 years of his time; it's now running twice as fast as his clock.

As the traveler returns home, she sees the homebody's clock advance from 4 to 20 years in 8 years of his time. She also sees her brother's clock advancing at twice the speed of her own. They both agree, however, that at the end of the trip the traveler's clock reads 16 years and the homebody's reads 20 years. Therefore, the traveler is 4 years younger.

The asymmetry in the paradox is that the traveler leaves Earth's reference frame and comes back, whereas the homebody never leaves Earth. It's also an asymmetry that the traveler and the homebody agree with the reading on the traveler's clock at each event but that they don't agree about the reading on the homebody's clock at each event. So, the traveler's actions define the events.

The Doppler effect and relativity together explain this effect mathematically at any instant. Note, too, that the speed at which an observed clock appears to run depends on whether it's traveling away from or toward the observer.

The fundamentals of this explanation have been exhaustively confirmed experimentally. In one such experiment, the lifetime of muon decay verifies the existence of time dilation. Stationary muons have a lifetime of about 2.2 microseconds. When traveling past an observer at $0.9994\ c$, muons' lifetime stretches to 63.5 microseconds, just as predicted by special relativity.

Experiments in which atomic clocks are transported at varying speeds have also produced results that confirm both special relativity and the twin paradox. In the Hafele-Keating experiment of 1971, for example, researchers flew cesium-beam atomic clocks around the world onboard commercial airliners—first eastbound and then westbound—and compared their times with stationary clocks at the US Naval Observatory. What this means is that the twin paradox today is more than just a theory.

6

A CHRONICLE OF TIMEKEEPING

Lesson 6 | A Chronicle of Timekeeping

Today, highly accurate timekeeping instruments set the beat for most of our electronic devices. Nearly all computers, for example, contain a quartz-crystal clock to regulate their operation. Such time-based technologies have become so integral to our daily lives that we recognize our dependency on them only when they fail to work. This lesson takes a look at their timekeeping predecessors and development.

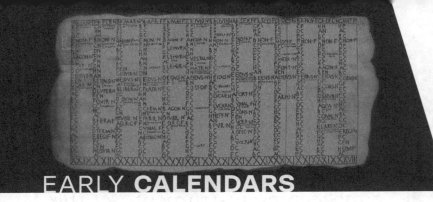

EARLY CALENDARS

- According to archaeological evidence, at least 5,000 years ago, the Babylonians, Egyptians, and other early civilizations began introducing calendars to organize and coordinate communal activities and public events, to schedule the shipment of goods, and, in particular, to regulate cycles of planting and harvesting.

- They based their calendars on three natural cycles: the solar day, marked by the successive periods of light and darkness as the Earth rotates on its axis; the lunar month, following the phases of the Moon as it orbits the Earth; and the solar year, defined by the changing seasons that accompany our planet's revolution around the Sun.

- Before the invention of artificial light, the Moon had greater social impact. Especially for those living near the equator, its waxing and waning was more conspicuous than the passing of the seasons. For this reason, calendars developed at lower latitudes were influenced more by the lunar cycle than by the solar year.

- In more northern climes, however, where seasonal agriculture was important, the solar year became more crucial. As the Roman Empire expanded northward, it organized its calendar for the most part around the solar year. Today's Gregorian calendar derives from the Babylonian, Egyptian, Jewish, and Roman calendars.

Lesson 6 | A Chronicle of Timekeeping

◢ The Egyptians formulated a civil calendar having 12 months of 30 days, with five days added to approximate the solar year. Each period of 10 days was marked by the appearance of special constellations called decans. At the rise of the star Sirius just before sunrise, which occurred around the all-important annual flooding of the Nile, 12 decans could be seen spanning the heavens.

TEMPORAL HOURS

◢ The cosmic significance the Egyptians saw in the 12 decans led them to develop a system in which each interval of darkness (and later, each interval of daylight) was divided into a dozen equal parts. These periods became known as temporal hours because their duration varied according to the changing length of days and nights with the passing of the seasons. Summer hours were long, and winter ones were short; only at the spring and autumn equinoxes were the hours of daylight and darkness equal.

◢ Temporal hours were adopted by the Greeks and then the Romans (who spread them throughout Europe). They remained in use for more than 2,500 years.

◢ To track temporal hours during the day, inventors devised sundials. The water clock was then designed to measure temporal hours at night. Although these devices performed well enough around the Mediterranean, they couldn't always be depended on in the cloudy and often freezing weather of northern Europe.

water clocks

WEIGHT-DRIVEN MECHANICAL CLOCKS

- The earliest recorded weight-driven mechanical clock was installed in 1283 at Dunstable Priory in Bedfordshire, England. It's not surprising that the Roman Catholic Church should've played a major role in the invention and development of clock technology: The strict observance of prayer times by monastic orders demanded a more reliable instrument of time measurement.

- By 1300, artisans were building clocks for churches and cathedrals in France and Italy. Because these early devices indicated the time by striking a bell (thereby alerting the surrounding community to its daily duties), they took their name from the Latin meaning "bell," *clocca*.

- The revolutionary aspect of this new type of timekeeper was neither the descending weight that provided its motive force nor the gear wheels that transferred the power. Instead, it was the part called the escapement. This device controlled the wheels' rotation and transmitted the power required to maintain the motion of the oscillator, the part that regulated the speed at which the timekeeper operated. The inventor of the clock escapement is unknown.

verge and foliot escapement

COUNTING HOURS

- In the early 14th century, a number of systems evolved governing the counting of hours. The schemes that divided the day into 24 equal parts varied according to the start of the count: Italian hours began at sunset and Babylonian hours at sunrise. Astronomical hours began at midday and "great clock" hours (used for some large public clocks in Germany) at midnight.

- Eventually, these and competing systems were superseded by "small clock," or French, hours. These split the day, as we currently do, into two 12-hour periods commencing at midnight. Minutes and seconds derive from the sexagesimal partitions of the degree introduced by Babylonian astronomers.

PORTABLE CLOCKS

- By the 15th century, a growing number of clocks were being made for domestic use. Those who could afford the luxury of owning a clock found it convenient to have one that could be moved from place to place.

- Innovators achieved portability by replacing the weight with a coiled spring. But the tension of a spring is greater after it's wound. To overcome this problem, the fusee (from *fusus*, the Latin term for "spindle") was invented by an unknown mechanical genius probably between 1400 and 1450.

Lesson 6 | A Chronicle of Timekeeping

- This cone-shaped device was connected by a cord to the barrel housing the spring. When the clock was wound, drawing the cord from the barrel onto the fusee, the diminishing diameter of the spiral of the fusee compensated for the increasing pull of the spring. Thus, the fusee equalized the force of the spring on the wheels of the timekeeper.

- The fusee allowed for the development of the portable clock as well as the subsequent evolution of the pocket watch. Many high-grade, spring-driven timepieces, such as marine chronometers, continued to incorporate this device until after World War II.

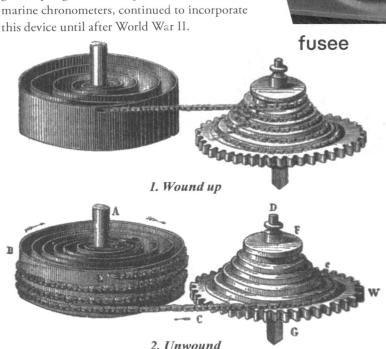

fusee

1. Wound up

2. Unwound

PENDULUM CLOCKS

- In the 16th century, Danish astronomer Tycho Brahe and his contemporaries tried to use clocks for scientific purposes. But even the best ones were still too unreliable. Astronomers in particular needed a better tool to time the transit of stars and create more accurate maps of the heavens.

- The pendulum proved to be the key to boosting the accuracy and dependability of timekeepers. The 27-year-old Dutch astronomer and mathematician Christiaan Huygens devised the first pendulum clock on Christmas Day in 1656.

- Pendulum clocks were about 100 times as accurate as their predecessors, reducing a typical gain or loss of 15 minutes a day to about a minute a week. Huygens immediately recognized both the commercial and scientific significance of his invention.

- Within six months, a local maker in the Hague had been granted a license to manufacture pendulum clocks. News of the invention spread rapidly, and by 1660, English and French artisans were developing their own versions of this new timekeeper.

Lesson 6 | A Chronicle of Timekeeping

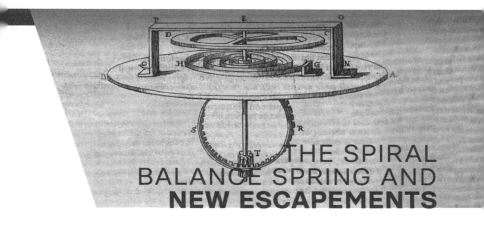

THE SPIRAL BALANCE SPRING AND NEW ESCAPEMENTS

◢ Huygens devoted much of his time to improving the device both for astronomical use and for solving the problem of finding longitude at sea. In 1675, he devised another fundamental improvement: the spiral balance spring.

◢ The spiral balance spring revolutionized the accuracy of watches, enabling them to keep time to within a minute a day. This advance sparked an almost immediate expansion of the market for watches, which were now typically carried in a pocket. This development also increased the demand for portable sundials by which watches could be set.

◢ At about the same time, Huygens heard news of an important English invention called the anchor escapement. Unlike the verge escapement he had been using in his pendulum clocks, the anchor escapement allowed the pendulum to swing in such a small arc that maintaining a cycloidal pathway became unnecessary.

◢ This escapement made the use of a long, seconds-beating pendulum more practical and thus led to the development of a new case design. The longcase clock began to emerge as one of the most popular English styles. Since 1876, these have been commonly called grandfather clocks (after a song by American Henry Clay Work). Longcase clocks with anchor escapements and long pendulums can keep time to within a few seconds a week.

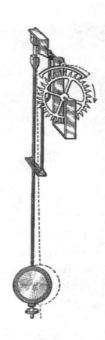

◢ The celebrated English clockmaker Thomas Tompion—and his successor, George Graham—later modified the anchor escapement to operate without recoil. This enhanced design, called the deadbeat escapement, became the most widely used type in precision timekeeping for the next 150 years.

anchor escapement

Lesson 6 | A Chronicle of Timekeeping

FINDING LONGITUDE

▲ When the Royal Observatory in Greenwich was founded in 1675, part of its charter was to find "the so-much-desired longitude of places." While navigators could determine their latitude at sea (their position north or south of the equator) by measuring the altitude of the sun or the polestar above the horizon, the heavens didn't provide such a straightforward solution for finding longitude.

▲ Storms and currents often confounded attempts to keep track of distance and direction traveled across oceans. The resulting navigational errors cost seafaring nations dearly, not just in prolonged voyages but also in loss of lives, ships, and cargo.

▲ The severity of this predicament was brought home to the British government in 1707, when an admiral of the fleet and more than 1,600 sailors perished in the wrecks of four Royal Navy ships off the coast of the Isles of Scilly. Thus, in 1714, through an act of Parliament, Britain offered substantial prizes for practical solutions to finding longitude at sea.

▲ The largest prize, £20,000, would be given to the inventor of an instrument that could determine a ship's longitude to within half a degree, or 30 nautical miles, when reckoned at the end of a voyage to a port in the West Indies, whose longitude could be accurately ascertained using proved land-based methods. The great reward attracted a deluge of harebrained schemes. Hence, the Board of Longitude, the committee appointed to review promising ideas, held no meetings for more than 20 years.

◢ In 1737, the Board of Longitude finally met for the first time to discuss the work of a most unlikely candidate, a Yorkshire carpenter named John Harrison. Harrison's large and rather cumbersome longitude timekeeper had been used on a voyage to Lisbon and on the return trip had proved its worth by correcting the navigator's dead reckoning of the ship's longitude by 68 miles.

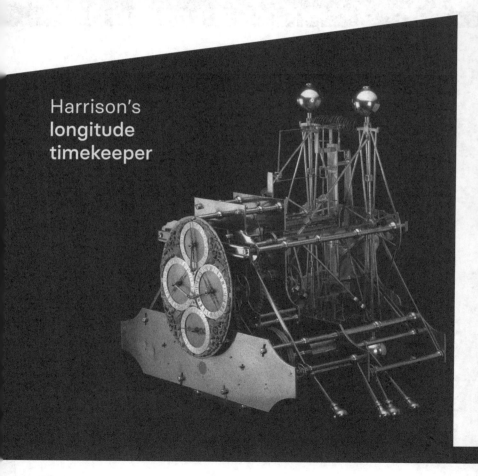

Harrison's **longitude timekeeper**

- But Harrison wasn't satisfied. Instead of asking the board for a West Indies trial, he requested and received financial support to construct an improved machine. After two years of work, still displeased with his second effort, Harrison embarked on a third design, laboring on it for 19 years. By the time it was ready for testing, he realized that his fourth marine timekeeper, a five-inch-diameter watch he had been developing simultaneously, was better.

- On a voyage to Jamaica in 1761, Harrison's oversize watch performed well enough to win the prize. But the board refused to give him his due without further proof. A second sea trial in 1764 confirmed his success. Harrison was reluctantly granted £10,000. Only when King George III intervened in 1773 did he receive the remaining prize money.

- Harrison's breakthrough inspired further developments. By 1790, the marine chronometer was so refined that its fundamental design never needed to be changed.

fourth attempt

THE TURN OF THE CENTURY

- At the turn of the 19th century, clocks and watches were relatively accurate, but they were still expensive. Recognizing the potential market for a low-cost timekeeper, two investors in Waterbury, Connecticut, took action.

- In 1807, they gave Eli Terry, a clockmaker in nearby Plymouth, a three-year contract to manufacture 4,000 longcase clock movements from wood. A substantial down payment allowed Terry to devote the first year to fabricating machinery for mass production. By manufacturing interchangeable parts, he completed the work within the terms of the contract.

- A few years later, Terry designed a wooden-movement shelf clock using the same volume-production techniques. For the relatively modest sum of $15, many average people could now afford a clock. This achievement helped establish what would become the renowned Connecticut clockmaking industry.

TIME ZONES

- Eventually, America's expanding railroad network required a uniform time standard for all the stations along the line. Astronomical observatories began distributing precise times to the railroad companies by telegraph.

- The first public time service, introduced in 1851, was based on clock beats wired from the Harvard College Observatory in Cambridge, Massachusetts. The Royal Observatory introduced its own time service the following year, creating a single standard time for Britain.

Lesson 6 | A Chronicle of Timekeeping

◢ The US established four time zones in 1883. By the next year, the governments of all nations had recognized the benefits of a worldwide standard of time for navigation and trade. At the 1884 International Meridian Conference in Washington DC, the globe was divided into 24 time zones. Delegates chose the Royal Observatory as the prime meridian (zero degrees longitude, the line from which all other longitudes are measured) in part because two-thirds of the world's shipping already used Greenwich time for navigation.

MASS PRODUCTION

◢ Many clockmakers of this era realized that the market for watches would far exceed that for clocks if production costs could be reduced. In Maine, a watchmaker named Aaron L. Dennison worried that American watchmakers seemed unable to compete against Europe, which controlled the market in the late 1840s.

◢ He met with Edward Howard, who had established a successful clock- and scale-making business in Roxbury, Massachusetts, to discuss mass-production methods for watches. Their efforts eventually led to the emergence of the American Waltham Watch Company.

◢ Their company benefited greatly from a huge demand for watches during the Civil War, when Union Army forces used them to synchronize operations. Improvements in fabrication techniques further boosted output and reduced prices significantly.

- Meanwhile, other US companies formed in the hope of capturing part of the burgeoning trade. Even some of the lower-grade American watches could keep reasonably good time. The watch was at last accessible to the masses. Later, self-winding mechanical wristwatches made their appearance during the 1920s.

LATER DEVELOPMENTS

Riefler clock

- At the end of the 19th century, Sigmund Riefler of Munich designed a radical new regulator—a highly accurate timekeeper that served as a standard for controlling others. Riefler's regulators attained an accuracy of a tenth of a second a day and were thus adopted by nearly every astronomical observatory.

- In 1928, Warren A. Marrison, an engineer at Bell Laboratories, then in New York City, discovered an extremely uniform and reliable frequency source that was as revolutionary for timekeeping as the pendulum had been 272 years earlier. Originally developed for use in radio broadcasting, the quartz crystal vibrates at a highly regular rate when excited by an electric current.

Lesson 6 | A Chronicle of Timekeeping

- The first quartz clocks installed at the Royal Observatory in 1939 varied by only two thousandths of a second a day. By the end of World War II, this accuracy had improved to the equivalent of a second every 30 years.

- Quartz-crystal technology didn't remain the premier frequency standard for long either, however. By 1948, Harold Lyons and his associates at the National Bureau of Standards in Washington DC had based the first atomic clock on a far more precise and stable source of timekeeping: an atom's natural resonant frequency, the periodic oscillation between two of its energy states.

- Subsequent experiments in both the US and England in the 1950s led to the development of the cesium-beam atomic clock. Today, the averaged times of cesium clocks in various parts of the world provide the standard frequency for Coordinated Universal Time, which has an accuracy of better than one nanosecond a day.

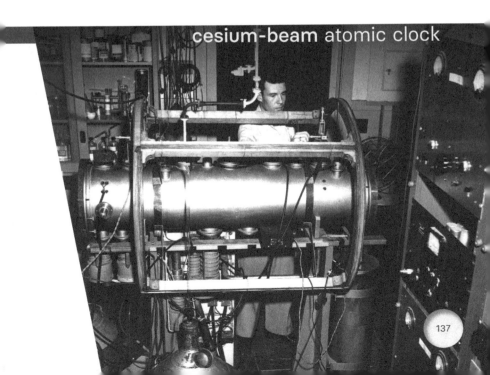

cesium-beam atomic clock

ABOUT THIS LESSON

This lesson was adapted from the article "A Chronicle of Timekeeping" by William J. H. Andrewes, a specialist in the field of time measurement who has worked at the Royal Observatory Greenwich, The Time Museum, and Harvard University.

Lesson 6 Transcript

A CHRONICLE OF TIMEKEEPING

This lesson was adapted from an article by William J. H. Andrewes, a specialist in the field of time measurement who has worked at the Royal Observatory Greenwich, The Time Museum and Harvard University.

Humankind's efforts to tell time have helped drive the evolution of our technology and science throughout history. The need to gauge the divisions of the day and night led the ancient Egyptians, Chinese, Greeks and Romans to create sundials, water clocks and other early chronometric tools. By the 13th century, demand for a dependable timekeeping instrument led medieval artisans to invent the mechanical clock.

This new device became accurate and reliable enough for scientific application when the pendulum was added. The precision timekeepers that were later developed resolved the critical problem of finding a ship's position at sea and went on to play key roles in the Industrial Revolution and the advance of modern civilization.

Today, highly accurate timekeeping instruments set the beat for most of our electronic devices. Nearly all computers, for example, contain a quartz-crystal clock to regulate their operation. Time signals beamed down from Global Positioning System satellites calibrate the functions not only of precision navigation equipment but also of cell phones, instant stock-trading systems and power-distribution grids. These time-based technologies have become so integral to our daily lives that we recognize our dependency on them only when they fail to work.

According to archaeological evidence, at least 5,000 years ago the Babylonians, Egyptians and other early civilizations began introducing calendars to organize and coordinate communal activities and public events, to schedule the shipment of goods and, in particular, to regulate cycles of planting and harvesting. They based their calendars on three natural cycles: the solar day, which is marked by the successive periods of light and darkness as the Earth rotates on its axis; the lunar month, following the phases of the Moon as it orbits the Earth; and the solar year, defined by the changing seasons that accompany our planet's revolution around the Sun.

Before the invention of artificial light, the Moon had greater social impact. Especially for those living near the equator, its waxing and waning was more conspicuous than the passing of the seasons. For this reason, calendars developed at lower latitudes were influenced more by the lunar cycle than by the solar year. In more northern climes, however, where seasonal agriculture was important, the solar year became more crucial. As the Roman Empire expanded northward, it organized its calendar for the most part around the solar year. Today's Gregorian calendar derives from the Babylonian, Egyptian, Jewish and Roman calendars.

The Egyptians formulated a civil calendar having 12 months of 30 days, with five days added to approximate the solar year. Each period of 10 days was marked by the appearance of special constellations which were called decans. At the rise of the star Sirius just before sunrise, which occurred around the all-important annual flooding of the Nile, 12 decans could be seen spanning the heavens.

The cosmic significance the Egyptians saw in the 12 decans led them to develop a system in which each interval of darkness (and later, also each interval of daylight) was divided into a dozen equal parts. These periods became known as temporal hours because their duration varied according to the changing length of days and nights with the passing of the seasons. Summer hours were long, winter ones were short; only at the spring and autumn equinoxes were the hours of daylight and darkness equal. Temporal hours were adopted by the Greeks and then by the Romans (who spread them throughout Europe) and remained in use for more than 2,500 years.

To track temporal hours during the day, inventors devised sundials, which indicate time by the length or direction of the Sun's shadow. The water clock was then designed to measure temporal hours at night. One of the first water clocks was simply a basin with a small hole near the bottom to allow the water to drip out. The falling water level denoted the passage of time as it dipped below hour lines inscribed on the inner surface.

Although these devices performed well enough around the Mediterranean, they couldn't always be depended on in the cloudy and often freezing weather of northern Europe.

The earliest recorded weight-driven mechanical clock was installed in 1283 at Dunstable Priory in Bedfordshire, England. It's not surprising that the Roman Catholic Church should've played a major role in the invention and development of clock technology. The strict observance of prayer times by monastic orders demanded a more reliable instrument of time measurement.

Plus, the Church not only controlled education but could also employ the most skillful craftsmen. And the growth of urban mercantile populations in Europe during the second half of the 13th century increased demand for improved timekeeping devices. By 1300 artisans were building clocks for churches and cathedrals in France and Italy and because these early devices indicated the time by striking a bell (and thereby alerting the surrounding community to its daily duties), they took their name from the Latin word for bell, *clocca*.

The revolutionary aspect of this new type of timekeeper was neither the descending weight that provided its motive force nor the gear wheels that transferred the power (which those had been around for at least 1,300 years); it was the part called the escapement. This device controlled the wheel's rotation and transmitted the power required to maintain the motion of the oscillator, the part that regulated the speed at which the timekeeper operated. The inventor of the clock escapement is unknown.

The mechanical clock was naturally suited to keeping equal hours rather than temporal hours that mark a fraction of a daytime or nighttime. But then using uniform hours raised the question of when to begin counting them. And so, in the early 14th century, a number of systems evolved. The schemes that divided the day into 24 equal parts varied according to the start of the count. Italian hours began at sunset and Babylonian hours at sunrise. Astronomical hours began at midday and "great clock" hours (which were used for some large public clocks in Germany) at midnight. Eventually these and competing systems were superseded by "small clock," or French, hours. These split the day, as we currently do, into two 12-hour periods commencing at midnight.

Minutes and seconds derive from the sexagesimal partitions of the degree introduced by Babylonian astronomers. The word *minute* has its origins in the Latin *prima minuta*, the first small division; the *second* comes from *secunda minuta*, the second small division.

So, the sectioning of the day into 24 hours and of hours and minutes into 60 parts each became so well established in Western culture that all efforts to change this arrangement have failed. The most notable attempt took place in revolutionary France in the 1790s, when the government adopted the decimal system. Although the French successfully introduced the meter, liter and other base-10 measures, the bid to break the day into 10 hours, each consisting of 100 minutes split into 100 seconds, lasted only 16 months.

For centuries after the invention of the mechanical clock, the periodic tolling of the bell in the town church or clock tower was enough to demarcate the day for most people. But by the 15th century, a growing number of clocks were being made for domestic use. Those who could afford the luxury of owning a clock found it convenient to have one that could be moved from place to place. Innovators achieved portability by replacing the weight with a coiled spring.

Now, the tension of a spring is greater after it's wound. To overcome this problem, the fusee (from *fusus*, the Latin term for *spindle*) was invented by an unknown mechanical genius probably between 1400 and 1450. This cone-shaped device was connected by a cord to the barrel housing the spring; when the clock was wound, drawing the cord from the barrel onto the fusee, the diminishing diameter of the spiral of the fusee compensated for the increasing pull of the spring. Thus, the fusee equalized the force of the spring on the wheels of the timekeeper.

The importance of the fusee shouldn't be underestimated: it allowed for the development of the portable clock as well as the subsequent evolution of the pocket watch. Many high-grade, spring-driven timepieces, such as marine chronometers, continued to incorporate this device until after World War II.

In the 16th century, Danish astronomer Tycho Brahe and his contemporaries tried to use clocks for scientific purposes. But even the best ones were just still too unreliable. Astronomers in particular needed a better tool to time the transit of stars and to create more accurate maps of the heavens. The pendulum proved to be the key to boosting the accuracy and dependability of timekeepers. The Italian physicist and astronomer Galileo, and others before him, had experimented with pendulums. But it was the 27-year-old Dutch astronomer and mathematician Christiaan Huygens who devised the first pendulum clock on Christmas Day in 1656.

Huygens saw that a pendulum traversing a circular arc completed small oscillations faster than large ones. Therefore, any variation in the extent of the pendulum's swing would cause the clock to gain or lose time. Realizing that maintaining a constant amplitude (or amount of travel) from swing to swing was impossible, Huygens devised a pendulum suspension that caused the bob to move in a cycloid-shaped arc rather than a circular one. Now, in theory, this enabled it to oscillate in the same time regardless of its amplitude.

Pendulum clocks were about 100 times as accurate as their predecessors, reducing a typical gain or loss of 15 minutes a day to instead about a minute a week. Huygens immediately recognized both the commercial and scientific significance of this invention and within six months a local maker in the Hague had been granted a license to manufacture pendulum clocks. News of the invention spread rapidly and by 1660 English and French artisans were developing their own versions of this new timekeeper.

The advent of the pendulum not only heightened demand for clocks but also resulted in their development as furniture. National styles soon emerged: English makers designed the case to fit around the clock movement. In contrast, the French placed greater emphasis on the shape and decoration of the case. Huygens, however, had little interest in these fashions, devoting much of his time to improving the device both for astronomical use and for solving the problem of finding longitude at sea. In 1675, he devised another fundamental improvement, the spiral balance spring.

Just as gravity controls the swinging oscillation of a pendulum in clocks, this spring regulates the rotary oscillation of a balance wheel in portable timepieces. A balance wheel is a finely balanced disk that rotates fully one way and then the other, repeating the cycle over and over. The spiral balance spring revolutionized the accuracy of watches, enabling them to keep time to within a minute a day. This advance sparked an almost immediate expansion of the market for watches, which were now no longer typically worn on a chain around the neck but instead carried in a pocket. This development also increased the demand for portable sundials by which the watches could be set.

At about the same time, Huygens heard news of an important English invention called the anchor escapement. Unlike the verge escapement he'd been using in his pendulum clocks, the anchor escapement allowed the pendulum to swing in such a small arc that maintaining a cycloidal pathway became unnecessary. This escapement made the use of a long, seconds-beating pendulum more practical, and thus led to the development of a new case design. The longcase clock began to emerge as one of the most popular English styles, and since 1876, these have been commonly called grandfather clocks (named after a song by American Henry Clay Work). Longcase clocks with anchor escapements and long pendulums can keep time to within a few seconds a week.

The celebrated English clockmaker Thomas Tompion—and his successor, George Graham—later modified the anchor escapement to operate without recoil. This enhanced design, called the deadbeat escapement, became the most widely used type in precision timekeeping for the next 150 years.

When the Royal Observatory in Greenwich was founded in 1675, part of its charter was to find "the so-much-desired longitude of places." The first Astronomer Royal, John Flamsteed, used clocks fitted with anchor and deadbeat escapements to time the exact moments that stars crossed the celestial meridian, an imaginary line that connects the poles of the celestial sphere and defines the due-south point in the night sky. This allowed him to gather more accurate information on star positions than had been possible by making angular measurements with sextants or quadrants alone.

While navigators could determine their latitude at sea (their position north or south of the equator) by measuring the altitude of the sun or the polestar above the horizon, the heavens didn't provide such a straightforward solution for finding longitude. Storms and currents often confounded attempts to keep track of distance and directions traveled across oceans and the resulting navigational errors cost seafaring nations dearly, not just in prolonged voyages, but also in loss of lives, ships and cargo.

The severity of this predicament was brought home to the British government in 1707, when an admiral of the fleet and more than 1,600 sailors perished in the wrecks of four Royal Navy ships off the coast of the Isles of Scilly. Thus, in 1714, through an act of Parliament, Britain offered substantial prizes for practical solutions to finding longitude at sea. The largest prize was £20,000 (which was at the time roughly 200 times the annual wage of a skilled engineer). The prize would be given to the inventor of an instrument that could determine a ship's longitude to within half a degree, or about 30 nautical miles, when reckoned at the end of a voyage to a port in the West Indies, whose longitude could be accurately ascertained using proved land-based methods.

The great reward attracted a deluge of harebrained schemes. Hence, the Board of Longitude, the committee appointed to review promising ideas, held no meetings for more than 20 years. Two approaches, though, had long been known to be theoretically sound. The first, called the lunar-distance method, involved precise observations of the Moon's position in relation to the stars to determine the time at a reference point from which longitude could be measured. The second required a very accurate clock to make the same determination.

Since the earth rotates once every 24 hours, or 15 degrees per hour, a two-hour time difference represents a 30-degree difference in longitude. The seemingly overwhelming obstacles to keeping accurate time at sea—such as the often violent motion of the ship or extreme changes in temperature and variations in gravity at different latitudes—led British physicist Isaac Newton and his followers to believe that the lunar-distance method, though problematic, was really the only viable solution.

Newton was wrong, however. In 1737, the Board of Longitude finally met for the first time to discuss the work of a most unlikely candidate, a Yorkshire carpenter named John Harrison. Harrison's large and rather cumbersome longitude timekeeper had been used on a voyage to Lisbon and on the return trip had proved its worth by correcting the navigator's dead reckoning of the ship's longitude by 68 miles. But Harrison wasn't satisfied. Instead of asking the board for a West Indies trial, he requested and received financial support to construct an improved machine.

After two years of work, still displeased with his second effort, Harrison embarked on a third design, laboring on it for 19 years. But by the time it was ready for testing, he realized that his fourth marine timekeeper, a five-inch-diameter watch he'd been developing simultaneously, was better.

On a voyage to Jamaica in 1761, Harrison's oversize watch performed well enough to win the prize. But the board refused to give him his due without further proof. A second sea trial in 1764 confirmed his success. Harrison was reluctantly granted £10,000. Only when King George III intervened in 1773 did he receive the remaining prize money. Harrison's breakthrough inspired further developments. By 1790, the marine chronometer was so refined that its fundamental design never needed to be changed.

At the turn of the 19th century, clocks and watches were relatively accurate, but they were still expensive. Recognizing the potential market for a low-cost timekeeper, two investors in Waterbury, Connecticut, took action. In 1807, they gave Eli Terry, a clockmaker in nearby Plymouth, a three-year contract to manufacture 4,000 longcase clock movements from wood. A substantial down payment allowed Terry to devote the first year to fabricating machinery for mass production. By manufacturing interchangeable parts, he completed the work within the terms of the contract.

A few years later, Terry designed a wooden-movement shelf clock using the same volume-production techniques. Unlike the longcase design, which required the buyer to purchase a case separately, Terry's shelf clock was completely self-contained. The customer could simply place it on a level shelf and wind it up. For the relatively modest sum of $15, many average people could now afford a clock. This achievement helped establish what would become the renowned Connecticut clockmaking industry.

Before the expansion of the railroads in the 19th century, towns across the US and Europe used the sun to determine local time. For example, since noon occurs in Boston about three minutes before it does in Worcester, Massachusetts, Boston's clocks were set about three minutes ahead of those in Worcester. The expanding railroad network, however, required a uniform time standard for all the stations along the same line.

Astronomical observatories began distributing precise times to the railroad companies by telegraph. The first public time service, introduced in 1851, was based on clock beats wired from the Harvard College Observatory in Cambridge, Massachusetts. The Royal Observatory introduced its own time service the following year, creating a single standard time for Britain.

The US established four time zones in 1883. By the next year, the governments of all nations had recognized the benefits of a worldwide standard time for navigation and trade. At the 1884 International Meridian Conference in Washington, DC, the globe was divided into 24 time zones. Delegates chose the Royal Observatory as the prime meridian (which was zero degrees longitude, the line from which all other longitudes are measured) in part because two thirds of the world's shipping already used Greenwich mean time for navigation.

Many clockmakers of this era realized that the market for watches would far exceed that for clocks if production costs could be reduced. But the problem of mass-fabricating interchangeable parts for watches was considerably more complicated because the precision demanded in making the miniaturized components was so much greater. Although improvements in quantity manufacture had been instituted in Europe since the late 18th century, European watchmakers' fears of saturating the market and threatening their workers' jobs by abandoning traditional practices stifled most thoughts of introducing machinery for the production of interchangeable watch parts.

In Maine, a watchmaker named Aaron L. Dennison worried that American watchmakers seemed unable to compete against Europe, which controlled the market in the late 1840s. So he met with Edward Howard, who'd established a successful clock- and scale-making business in Roxbury, Massachusetts, to discuss mass-production methods for watches. Howard and his partner gave Dennison space to experiment and develop machinery for the project. By the fall of 1852, 20 watches had been completed. Dennison's workmen finished 100 watches by the following spring, and 1,000 more a year later. By that time, the manufacturing facilities in Roxbury were proving too small, so the newly named Boston Watch Company moved to Waltham, Massachusetts. By the end of 1854, it was assembling 36 watches every week.

The American Waltham Watch Company, as it eventually became known, benefited greatly from a huge demand for watches during the Civil War, when Union Army forces used them to synchronize operations. Improvements in fabrication techniques further boosted output and reduced the prices significantly.

Meanwhile, other US companies formed in the hope of capturing part of this burgeoning trade. The Swiss, who'd previously dominated the industry, grew concerned when their exports plummeted in the 1870s. The investigator they sent to Massachusetts who discovered that not only was productivity higher at the Waltham factory but production costs were less. Even some of the lower-grade American watches could keep reasonably good time. The watch was at last accessible to the masses.

Because women had worn bracelet watches in the 19th century, wristwatches were long considered feminine accoutrements. But during World War I, the pocket watch was modified so that it could be strapped to the wrist, where it could be viewed more readily on the battlefield. With the help of a substantial marketing campaign, the masculine fashion for wristwatches caught on after the war. Self-winding mechanical wristwatches made their appearance during the 1920s.

At the end of the 19th century, Sigmund Riefler of Munich designed a radical new regulator, a highly accurate timekeeper that served as a standard for controlling others. Housed in a partial vacuum to minimize the effects of barometric pressure and equipped with a pendulum that was largely unaffected by temperature variations, Riefler's regulators attained an accuracy of a tenth of a second a day and were thus adopted by nearly every astronomical observatory.

In 1928, Warren A. Marrison, an engineer at Bell Laboratories, which was then in New York City, discovered an extremely uniform and reliable frequency source that was as revolutionary for timekeeping as the pendulum had been 272 years earlier. Originally developed for use in radio broadcasting, the quartz crystal vibrates at a highly regular rate when excited by an electric current. The first quartz clocks installed at the Royal Observatory in 1939 varied by only 2,000ths of a second every day. By the end of World War II, this accuracy had improved to the equivalent of a second every 30 years.

Quartz-crystal technology didn't remain the premier frequency standard for long either, however. By 1948, Harold Lyons and his associates at the National Bureau of Standards in Washington, DC, had based the first atomic clock on a far more precise and stable source of timekeeping: an atom's natural resonant frequency, the periodic oscillation between two of its energy states. Subsequent experiments in both the US and England in the 1950s led to the development of the cesium-beam atomic clock. Today, the averaged times of cesium clocks in various parts of the world provide the standard frequency for Coordinated Universal Time, which has an accuracy of better than one nanosecond a day.

Until the mid-20th century, the sidereal day, the period of the Earth's rotation on its axis in relation to the stars, was used to determine standard time. This practice was retained even though it had been suspected ever since the late 18th century that our planet's axial rotation isn't entirely constant. But the rise of cesium clocks capable of measuring discrepancies in the Earth's spin meant a change was necessary. A new definition of the second, based on the resonant frequency of the cesium atom, was adopted as the new standard unit of time in 1967.

The precise measurement of time is so fundamental to science and technology that the search for ever greater accuracy continues. For about 50 years, the performance of atomic clocks had been improving by a factor of at least 10 per decade. But over the past decade improvements in atomic clock accuracy have dramatically accelerated.

Recent advances in atomic physics and laser science—particularly the Nobel Prize–winning development of femtosecond laser frequency combs—have enabled the development of many new types of optical atomic clocks. Some are based on transitions in single ions in electromagnetic traps while others are based on collections of cold neutral atoms held in lattices formed by laser light. Several of these atomic clocks are already stable to within a few hundred femtoseconds per day and they continue to improve rapidly.

At this level of performance, formerly negligible effects become important and measurable. For example, the best atomic clocks can now measure changes in gravity over the distance of a stair step, tiny magnetic fields generated by heart and brain activity, and other quantities such as temperature and acceleration.

Companies are now manufacturing "chip-scale" atomic clocks the size of a quarter. In addition to keeping time with increasing accuracy, new generations of atomic clocks will be used as exquisite sensors for myriad applications and will become ever smaller and ever more portable.

Although our clocks will surely improve in the future, nothing will ever change the fact that if there's one thing we never have enough of, it's time.

7
ATOMIC CLOCKS

Einstein's general theory of relativity predicts that clocks experiencing different gravitational pulls will tick at different rates. A clock at higher elevation will tick faster than will a clock closer to Earth's center. Einstein's special theory of relativity predicts a similar effect for clocks in motion. A stationary clock will tick faster than a moving clock. This lesson looks at efforts to observe those time dilation effects as well as advances in high-precision timekeeping.

Lesson 7 | Atomic Clocks

OBSERVING TIME DILATION

- Both of the aforementioned time dilation effects have been verified in a number of experiments throughout the decades, which have traditionally depended on large scales of distance or velocity. In one landmark test, in 1971, Joseph Hafele of Washington University in St. Louis and Richard Keating of the US Naval Observatory flew cesium atomic clocks around the world on commercial jet flights, then compared the clocks with reference clocks on the ground to find that they had diverged, as predicted by relativity.

- Yet even at the speed and altitude of jet aircraft, the effects of time dilation are tiny. In the Hafele-Keating experiment, the atomic clocks differed after their journeys by just tens to hundreds of nanoseconds.

- Thanks to improved timekeeping, similar demonstrations can now take place at more mundane scales in the laboratory. In a series of experiments described in *Science* magazine, researchers at the National Institute of Standards and Technology, or NIST, registered differences in the passage of time between two high-precision optical atomic clocks when one was elevated by just a third of a meter or when one was set in motion at speeds of less than 10 meters per second.

cesium atomic clock used in Hafele-Keating experiment

- Again, the effects are minuscule: It would take the elevated clock hundreds of millions of years to log one more second than its counterpart, and a clock moving a few meters per second would need to run about as long to lag one second behind the stationary clock. But the development of optical clocks based on aluminum ions, which can keep time to within one second in roughly 3.7 billion years, allows researchers to expose those tiny relativistic effects.

ULTIMATE CLOCKS

- Thanks to major technical advances, the art of ultraprecise timekeeping has been progressing with a speed not seen for decades. These days, a good cesium beam clock will tick off seconds true to about a microsecond a month, its frequency accurate to five parts in 10^{13}.

- The primary time standard for the US, a cesium fountain clock installed in 2014 by the NIST at its Boulder, Colorado, laboratory, is good to three parts in 10^{16}. That's 2,000 times the accuracy of NIST's best clock in 1975. Successful prototypes of new clock designs—devices that extract time from aluminum or mercury ions instead of cesium—have recently attained accuracy in the 10^{-18} power range, a 100-fold improvement in a decade.

- The second was defined in 1967 by international fiat to be "the duration of 9,192,631,770 periods of the radiation corresponding to the transition between the two hyperfine levels of the ground state of the cesium 133 atom." Under that standard, to measure a second, one has to look at cesium. The best clocks now don't, so, strictly speaking, they don't measure seconds. That's one predicament the clockmakers face.

THE ADVANTAGE OF ATOMIC CLOCKS

Every clock has at least two basic components: an oscillator and a counter. An atomic clock is so accurate because it includes a third element: a feedback system that periodically checks an atomic reference to keep the oscillator ticking with nearly perfect regularity. In a state-of-the-art optical ion clock, an ultraviolet probe laser serves as the oscillator. Pulses of infrared laser light yield a counter. One electron orbiting a single, nearly motionless mercury atom functions as the ultimate reference.

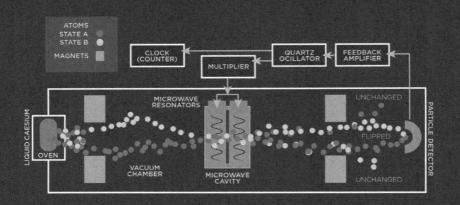

- Further down the road lies a more fundamental limitation: As Albert Einstein theorized and as experiment has confirmed, time isn't absolute. The rate of any clock slows down when gravity gets stronger or when the clock moves quickly relative to its observer. By putting ultraprecise clocks on the International Space Station, scientists hope to put relativity theory through its toughest tests yet.

- Clocks with a precision of 10^{-18} represent proportions that correspond to a deviation of less than half a second over the age of the universe. That means the effects of relativity are starting to test the scientists. No technology exists that can synchronize clocks around the world with such exactness.

FOUNTAIN CLOCKS

- The ticker in a cesium clock is neither mechanical (like a pendulum) nor electromechanical (like a quartz crystal). It's quantum-mechanical: A photon of light is absorbed by the cesium atom's outermost electron, causing the electron to flip its magnetic field upside down.

- All cesium atoms are identical. Each will flip the spin of its outer electron when hit with microwaves at the frequency of exactly 9,192,631,770 cycles per second. To measure seconds, the clock locks its microwave generator onto the spot in the spectrum where the most cesium atoms react. Then it starts counting cycles.

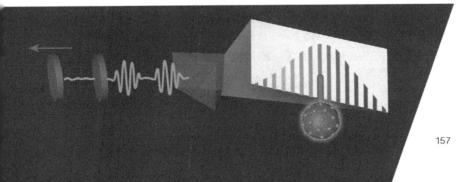

◢ Complicating matters is the Heisenberg indeterminacy principle, which puts strict limits on how precisely one can measure the frequency of a single photon. The best clocks now scan a one-hertz-wide sweet spot to find its exact center, plus or minus one millihertz, in every single measurement—despite the Heisenberg limits.

◢ "The reason we can do it is that we look at more than a million atoms each time," explained physicist Kurt Gibble. "Because it isn't really just one measurement, it doesn't violate the laws of quantum mechanics." However, that solution creates other problems.

◢ At room temperature, cesium is a soft, silvery metal. It would melt in a person's palm to a golden puddle, and it reacts violently with water.

◢ Inside a cesium beam clock, an oven heats the metal until atoms boil off. These hot particles can zip through the microwave cavity at various speeds and angles. Some move so fast that, because of relativity, they behave as if time has slowed. Because of Doppler shifting to other atoms, the microwaves appear to be higher or lower in frequency than they are.

◢ The atoms no longer behave identically, so the ticks grow less distinct. Heisenberg would probably have suggested slowing the atoms down, and that's what clockmakers have done. Several clocks around the world—at NIST; the US Naval Observatory; and the standards institutes in Paris; Teddington, England; and Brunswick, Germany—toss supercooled balls of cesium atoms in a fountainlike arc through a microwave.

◢ To condense the hot cesium gas into a ball, six intersecting laser beams decelerate the atoms to less than two microkelvins—almost a complete ultracold standstill. The low temperature all but eliminates relativistic and Doppler shifts, and it gives a two-meter-tall fountain clock half a second to flip the atoms' spins.

CLOCKS IN SPACE

- Introduced in 1996, fountain clocks rapidly knocked 90% off the uncertainty of international atomic time. However, fountain clocks still rush the job. "We would have to quadruple the height of the tower to double the observation time," said Donald Sullivan, former chief of the time and frequency division at NIST.

- Sullivan led one of three projects to put fountainlike clocks on the International Space Station. Per Sullivan, "In space, we can launch a ball of atoms at 15 centimeters per second through a 74-centimeter cavity." That provides "5 to 10 seconds to observe them," he explains.

- The $25-million Primary Atomic Reference Clock in Space (PARCS) project he worked on was designed to turn out seconds good to five parts in 10^{17}. PARCS was canceled in 2004, when NASA shifted funding from the space station.

- However, a device from the European Space Agency called ACES (Atomic Clock Ensemble in Space) is scheduled to launch in 2021. It aims to measure with 99.99997% accuracy how much the microgravity of low Earth orbit slows time compared with measurements made on the ground.

One of two atomic clocks developed for ACES mission.

RACE

- Efforts to make a third space-faring clock, called RACE (Rubidium Atomic Clock Experiment), helped to refine a newer approach that replaces cesium with a different alkali element. The project was cancelled along with PARCS, but the following is a look at how it could work. "In the best cesium fountains, the largest source of error are so-called cold collisions," explained Gibble, who directed the RACE project.

- At temperatures near absolute zero, quantum physics takes over, and atoms start to behave like waves. "They appear hundreds of times bigger than normal," Gibble continued, "so they collide much more often. At a microkelvin, cesium has nearly the maximum possible cross section."

- However, "the effective size for rubidium atoms is 50 times smaller." That enables rubidium clocks to reach one fifth the uncertainty of ACES. Rubidium clocks offer another advantage as well: the opportunity to look for fluctuations in the fine-structure constant, alpha (α).

- Alpha is a fundamental physical constant that determines the strength of electromagnetic interactions in atoms and molecules. Its value is very nearly $1/137$, a unitless number that falls out of the equations for the Standard Model of physics. It is a very important number—change α very much, and the universe couldn't support life as we know it.

- In the Standard Model, the fine-structure constant is immutable throughout eternity. However, in some competing theories (such as certain string theories), α could waver slightly or grow as time goes by.

ION CLOCKS

- The development of ion clocks in some ways has surpassed clocks based on fountains of atoms. In 2001, Scott A. Diddams and his colleagues at NIST reported an initial trial run of an optical atomic clock based on a single mercury atom.

- It may seem like a natural idea to graduate from microwaves, at frequencies of gigahertz, to visible light, well into the terahertz part of the spectrum. Optical photons pack enough energy to bump electrons clear into the next orbital shell, removing the need to deal with subtleties like spin. But although the ticker still works at terahertz frequencies, the counter breaks.

- The problem was that "nobody knows how to count 10 to the 16 power cycles per second," observed Eric A. Burt of NASA's Jet Propulsion Laboratory. "We needed a bridge to the microwave regime, where we do have electronic counters."

- That is where optical rulers come in. In 1999, researchers at the Max Planck Institute for Quantum Optics in Garching, Germany, figured out a way to measure optical frequencies directly, using a reference laser that pulses at a rate of one gigahertz.

- Each pulse of light is just a couple of dozen femtoseconds long. (A femtosecond is a very, very small amount of time. More femtoseconds elapse in each second than there have been hours since the big bang.)

- A laser puts out a continuous beam of only one color, but pulsing that laser produces a mixture of colors in each flash. The spectrum of a femtosecond pulse is a bizarre thing to see: millions of sharp lines spanning the rainbow, each line spaced exactly the same distance from its neighbors.

LOOKING FORWARD

- Diddams's group at NIST has built a rudimentary optical clockwork around mercury ions, which they immobilize in an electromagnetic trap. Because each atom is missing an electron, the ions carry a positive charge. They repel one another, so collisions are not a problem. The device is stable to better than six parts in 10^{16} over the course of a second.

- Groups at the Federal Institute of Physics and Metrology in Brunswick, Germany, and elsewhere are experimenting with uncharged calcium atoms. Because neutral atoms can be crammed more densely into the trap than ions can, the signal soars higher over the noise.

- In 2015, a NIST-led team reported the successful demonstration of an optical lattice clock based on atoms of strontium 87; the uncertainty was pegged at two parts in 10^{18}. A German research group reported in 2016 on a precision almost as good, of 3.2 parts in 10^{18}, from a system based on a single ion of ytterbium 171. The NIST group also used ytterbium for a lattice clock that achieved an accuracy of 1.6 parts in 10^{18}.

ytterbium lattice clock

◢ However, note that the word *accuracy* keeps appearing. These new clocks "move away from the atomic definition of the second, which is based on the properties of cesium," Sullivan points out. For the newest and best clocks to be strictly accurate as keepers of the time to which we set our watches, that definition will have to change.

ABOUT THIS LESSON

This lesson was adapted from the articles "How Time Flies" by John Matson, a former reporter and editor for *Scientific American* who has written extensively about astronomy and physics, and "Ultimate Clocks" by W. Wayt Gibbs, a contributing editor for *Scientific American*.

Lesson 7 Transcript

ATOMIC CLOCKS

This lesson was adapted from the articles "How Time Flies" by John Matson, a former reporter and editor for *Scientific American* who has written extensively about astronomy and physics, and the article "Ultimate Clocks" by W. Wayt Gibbs, a contributing editor for *Scientific American*.

If you've ever found yourself cursing noisy upstairs neighbors, take solace in the fact that they are aging faster than you are. Einstein's general theory of relativity predicts that clocks experiencing different gravitational pulls will tick at different rates. So, a clock at higher elevation will tick faster than a clock that is closer to Earth's center. In other words, time passes more quickly in your neighbor's upstairs apartment than it does in yours.

Einstein's special theory of relativity predicts a similar effect for clocks that are in motion—a stationary clock will tick faster than a moving clock. This is the source of the famous twin paradox: Following a roundtrip journey on a spaceship traveling at some exceptionally high velocity, a traveler would return to Earth to find that her twin sibling is now older than she is, because time has passed more slowly on her moving ship than on Earth.

So, both of these so-called time dilation effects have been verified in a number of experiments throughout the decades, which have traditionally depended on large scales of distance or velocity.

In one landmark test, in 1971, Joseph Hafele of Washington University in St. Louis and Richard Keating of the US Naval Observatory flew cesium atomic clocks around the world on commercial jet flights, then compared the clocks with reference clocks that had stayed on the ground, and they found that the clocks had diverged, which was exactly what was predicted by relativity. Yet even at the speed and altitude of jet aircraft, the effects of time dilation are tiny; in the Hafele-Keating experiment, the atomic clocks differed after their journeys by just tens to hundreds of nanoseconds.

Thanks to improved timekeeping, similar demonstrations can now take place at more mundane scales in the laboratory. In a series of experiments described in *Science* magazine, researchers at the National Institute of Standards and Technology, or NIST, registered differences in the passage of time between two extremely high-precision optical atomic clocks when one was elevated by just a third of a meter or when one was set in motion at speeds of less than 10 meters per second.

So, again, the effects are minuscule. It would take the elevated clock hundreds of millions of years to log one more second than its lower counterpart and a clock moving at a few meters per second would need to run about as long to lag one second behind the stationary clock. But the development of optical clocks based on aluminum ions, which can keep time to within one second in roughly 3.7 billion years, allows researchers to expose these very tiny relativistic effects.

"People usually think of it as negligible, but for us it is not," said study leader James Chin-wen Chou of NIST. "We can definitely see it," he said.

The NIST group's optical clocks use lasers to probe the quantum state of aluminum ions held in radio-frequency traps. When the laser's frequency is just right, it resonates with a transition between quantum states in the aluminum ion that has a constant frequency. By continually tuning the laser to drive that aluminum transition, an interaction that occurs only in a tiny window near 1.121 petahertz, which is about 1.121 quadrillion cycles per second, the laser's frequency can be stabilized to an exquisitely sensitive degree, allowing it to act as the clock's pendulum.

"If we anchor the frequency of the oscillator—in our case, laser light—to the unchanging, stable optical transition in aluminum, the laser oscillation can serve as the tick of the clock," Chou explains. To put the sensitivity of the optical clocks in perspective, Chou notes that the two timekeepers in the study differed after a height change of a mere step on a staircase—never mind the entire floor separating you from your noisy neighbors—or with just a few meters per second of motion. "If you push your daughter on a swing," he says, "it's about that speed."

In the past, such relativistic experiments have involved either massive scales of distance or velocity or else oscillations so fast that their ticks cannot be reliably counted for timing purposes, said Holger Müller, an atomic physicist at the University of California, Berkeley. "It's an enormous achievement that you can build optical clocks so good that you can now see relativity in the lab," he said.

Müller has used atom interferometry to make precision measurements of relativistic effects—measurements that rely not on counting individual oscillations but rather on tracking the interference between quantum waves representing individual atoms. It is a process akin to striking two tuning forks to listen to the pulsations of their interference without actually knowing or measuring how many times each fork individually vibrates.

The study at NIST "operates on familiar scales of distance and velocity, with clocks that can be used for universal timing applications," Müller says. "They see the effects of general and special relativity, and that makes relativity something you can kind of see and touch."

One muggy week in May 2002, dozens of the top clockmakers in the world convened in New Orleans to present their latest inventions. They were all scientists, and their conversations buzzed with talk of spectrums and quantum levels, rather than gears and escapements. Today anyone who wants to build a more accurate clock must advance into the frontiers of physics and engineering in several directions at once. They're cobbling lasers that spit out pulses a quadrillionth of a second long together with chambers that chill atoms to a few millionths of a degree above absolute zero. They're snaring individual ions in tar pits of light and magnetism and manipulating the spin of electrons in their orbits.

Thanks to major technical advances, the art of ultraprecise timekeeping has been progressing with a speed not seen for decades. These days a good cesium beam clock will tick off seconds true to about a microsecond a month, its frequency accurate to five parts in 10 to the 13th power.

The primary time standard for the United States, a cesium fountain clock installed in 2014 by the National Institute of Standards and Technology (NIST) at its Boulder, Colorado laboratory, is good to three parts in 10 to the 16th power. That's 2,000 times the accuracy of NIST's best clock in 1975. Successful prototypes of new clock designs—devices that extract time from aluminum or mercury ions instead of cesium—have recently attained accuracy in the 10 to the 18th power range, a 100-fold improvement in a decade.

Accuracy may not be quite the right word. The second was defined in 1967 by the international fiat to be technically "the duration of 9,192,631,770 periods of the radiation corresponding to the transition between the two hyperfine levels of the ground state of the cesium 133 atom." Leave aside for the moment what that means: the point is that to measure a second, you have to look at cesium. The best clocks now don't. So, strictly speaking, they don't measure seconds. That's one predicament the clockmakers face.

Further down the road lies a more fundamental limitation. As Einstein theorized and as experiment has confirmed, time is not absolute. The rate of any clock slows down when gravity gets stronger or when the clock moves quickly relative to its observer—even a single photon emitted as an electron reorients its magnetic poles or jumps from one orbit to another. By putting ultraprecise clocks on the International Space Station, scientists hope to put relativity theory through its toughest tests yet.

Now clocks have achieved a precision of 10 to the 18th power—proportions that correspond to a deviation of less than half a second over the age of the entire universe. That means the effects of relativity are starting to test the scientists. No technology exists that can synchronize clocks around the world with such exactness.

So, in that case, why bother to improve atomic clocks? The duration of the second can already be measured to 14 decimal places, a precision 1,000 times that of any other fundamental unit. One reason to do better is that the second is increasingly the fundamental unit. Four of the six other basic units—including the meter, lumen, ampere, and most recently the kilogram—are now defined in terms of the second. By improving clocks, scientists can improve measurements of much more than time.

More stable and portable clock designs could also be a big boon to navigation, enhancing the accuracy and reliability of the Global Positioning System and of Galileo, a competing system in Europe. Better clocks would help NASA track its satellites, enable utilities and communications firms to trace faults in their networks, and enhance geologists' ability to pinpoint earthquakes and nuclear bomb tests. Astronomers could use them to connect telescopes in ways that dramatically sharpen their images. Inexpensive, microchip-size atomic clocks are likely to have myriad uses that we have not yet imagined.

To understand why timekeeping has suddenly lurched into high gear, it helps to know a little about how atomic clocks work. Every clock has at least two basic components, an oscillator and a counter. An atomic clock is so accurate because it includes a third element— a feedback system that periodically checks an atomic reference to keep the oscillator ticking with near perfect regularity.

In a state-of-the-art optical ion clock, an ultraviolet probe laser serves as the oscillator. Pulses of infrared laser light yield a counter. One electron orbiting a single, nearly motionless mercury atom functions as the ultimate reference. The atom, boiled off a piece of mercury in an oven, is ionized when a current strips away one of its electrons, leaving it with a positive charge. An electromagnetic field then confines the ion to the center of a ring-shaped trap.

The beam of a so-called cooling laser causes the ion's outermost electron to jump millions of times a second to a higher, unstable orbit, fluorescing each time it falls back to the ground level. The fluorescence has two functions: It cools the atom to nearly absolute zero, and it allows scientists to verify (through a microscope) that the clock is still running. Once the atom is cool, stable and glowing, it's ready to serve as the clock's reference.

The closest thing to a ticker in an ion clock is the probe laser. The color of the photons streaming from the laser reflects the frequency of their oscillation. To check that their frequency hasn't slowed or quickened, the laser periodically shines on the mercury atom and scientists tune the color of the probe light to the precise frequency that knocks the ion's outer electron into a metastable orbit, thus shelving the electron for up to half a second. When the laser is tuned to this special frequency, the electron stops fluorescing, and the ion goes dark. If the laser oscillator drifts, the ion blinks back on.

A feedback system adjusts the laser color until the fluorescence is at a minimum and the probe light, now rock steady, is next passed via optical fiber to a counter. The probe light oscillates about a quadrillion times a second, far too fast to count directly. So a third laser acts like a reducing gear to translate the time signal from a quadrillion cycles a second to about a billion cycles a second. This third laser emits infrared pulses just a few femtoseconds long, with stretches of darkness between them.

The trick is to lock its pulse rate in perfect synchronicity with the frequency of the probe light. To do this, the clockwork exploits a curious fact: when passed through a prism, each ultrashort pulse splits into a rainbow of colors spaced at regular frequency intervals, like the teeth on a gear.

By moving an adjustable mirror, scientists alter the delay between the pulses, thereby stretching or compressing the range of frequencies carried by each pulse. This allows them to position the gear so that one of its teeth matches the color (and thus the frequency) of the probe light—which means it's also locked to the hardwired behavior of the mercury ion. An electronic detector then counts the synchronized pulses as they go by, a billion a second, ticking off the passage of time.

So, in principle, an atomic clock is just like any other timepiece, with an oscillator that ticks in a regular way and a counter that converts the ticks to seconds. The ticker in a cesium clock is neither mechanical (like a pendulum) nor electromechanical (like a quartz crystal). Instead, it's quantum-mechanical— a photon of light is absorbed by the cesium atom's outermost electron, causing the electron to flip its magnetic field upside down.

Unlike pendulums and crystals, all cesium atoms are identical. And every one will flip the spin of its outer electron when hit with microwaves at the frequency of exactly 9,192,631,770 cycles per second. To measure seconds, the clock locks its microwave generator onto the sweet spot in the spectrum where the most cesium atoms react. Then it starts counting cycles.

Of course, nothing in quantum physics is really that simple. Complicating things, as usual, is the Heisenberg indeterminacy principle, which puts strict limits on how precisely one can measure the frequency of a single photon. The best clocks now scan a one-hertz-wide sweet spot to find its exact center, plus or minus one millihertz, in every single measurement—despite the Heisenberg limits. "The reason we can do it is that we look at more than a million atoms each time," explained Kurt Gibble, a physicist at Pennsylvania State University, who was at the New Orleans conference. "Because it isn't really just one measurement, it doesn't violate the laws of quantum mechanics."

But that solution creates other problems. At room temperature, cesium is a soft, silvery metal. It would melt in your palm to a golden puddle—although you wouldn't want to touch it, because it reacts violently with water. Inside a cesium beam clock, an oven heats the metal until atoms boil off. These hot particles can zip through the microwave cavity at various speeds and angles. Some move so fast that, because of relativity, they behave as if time has slowed. Because of Doppler shifting to other atoms, the microwaves appear to be higher or lower in frequency than they actually are. The atoms no longer behave identically, so the ticks grow less distinct.

Heisenberg would probably have suggested slowing the atoms down and that's what clockmakers have done. Several clocks around the world—at NIST, the US Naval Observatory, and the standards institutes in Paris, Teddington, England, and Brunswick, Germany—toss supercooled balls of cesium atoms in a fountainlike arc through a microwave.

To condense the hot cesium gas into a ball, six intersecting laser beams decelerate the atoms to less than two microkelvins—almost a complete ultracold standstill. The low temperature all but eliminates relativistic and

Doppler shifts, and it gives a two-meter-tall fountain clock half a second to flip the atoms' spins. Introduced in 1996, fountain clocks rapidly knocked 90% off the uncertainty of international atomic time.

It takes time to make a good second, though, and the fountain clocks still rush the job. "We would have to quadruple the height of the tower to double the observation time," says Donald Sullivan, former chief of the time and frequency division at NIST. But instead of punching a hole through the ceiling of his lab, Sullivan led one of three projects to put fountainlike clocks in the International Space Station. "In space, we can launch a ball of atoms at 15 centimeters per second through a 74-centimeter cavity. So we have 5 to 10 seconds to observe them," he explains.

The $25-million Primary Atomic Reference Clock in Space (PARCS) project that he worked on was designed to turn out seconds good to five parts in 10 to the 17th power. PARCS was canceled in 2004, when NASA shifted funding from the space station to programs to send astronauts to the Moon and, perhaps eventually, Mars. But a device from the European Space Agency called ACES (Atomic Clock Ensemble in Space) is scheduled to launch in 2021 and aims to measure within 99.99997% accuracy how much the microgravity of low Earth orbit slows time compared with measurements that are made on the ground.

Efforts to make a third space-faring clock, which is called RACE (Rubidium Atomic Clock Experiment), helped to refine a newer approach that replaces cesium with a different alkali element. That project was cancelled along with PARCS, but here's how it could work.

"In the best cesium fountains the largest source of error are so-called cold collisions," explained Gibble, who directed the RACE project. At temperatures near absolute zero, quantum physics takes over and atoms start to behave like waves. "They appear hundreds of times bigger than normal," he said, "so they collide much more often. At a microkelvin, cesium has nearly the maximum possible cross section. But the effective size for rubidium atoms is 50 times smaller." That enables rubidium clocks to reach one fifth the uncertainty of ACES.

Rubidium clocks offer another advantage, as well: the opportunity to look for fluctuations in the fine-structure constant, which is referred to as *alpha* (α). *Alpha* is a fundamental physical constant that determines the strength of electromagnetic interactions in atoms and molecules. Its value is very nearly 1 divided by 137, a unitless number that falls out of the equations for the Standard Model of physics. It is a very important number—change *alpha* very much, and the universe would not support life as we know it.

In the Standard Model, the fine-structure constant is immutable throughout eternity. But in some competing theories (such as certain string theories), *alpha* could waver slightly or grow as time goes by. In 2001, a group of astronomers reported preliminary evidence that *alpha* may have increased by one part in 10,000 during the past six billion years. But the evidence is equivocal and the question is a hard one to settle. By analyzing radio signals from a distant galaxy, in 2018 astronomers concluded that if *alpha* is fluctuating, it has varied by no more than about one part in a million over the past 2.9 billion years.

The development of ion clocks in some ways has surpassed clocks based on fountains of atoms. In 2001, Scott A. Diddams and his colleagues at NIST reported an initial trial run of something many clock builders thought they would never see— an optical atomic clock based on a single mercury atom. It may seem like a natural idea to graduate from microwaves, which have frequencies of a gigahertz, to visible light, which well into the terahertz part of the spectrum. Optical photons pack enough energy to bump electrons clear into the next orbital shell—no need to fuss with subtleties like spin. But although the ticker still works at terahertz frequencies, the counter breaks.

The problem was that "nobody knows how to count 10 to the 16 power cycles per second," said Eric A. Burt of NASA's Jet Propulsion Laboratory. "We needed a bridge to the microwave regime, where we do have electronic counters." Enter the optical ruler. In 1999, researchers at the Max Planck Institute for Quantum

Optics in Garching, Germany, figured out a way to measure optical frequencies directly, using a reference laser that pulses at a rate of one gigahertz. Each pulse of light is just a couple of dozen femtoseconds long. (A femtosecond, by the way, is a very, very small amount of time. More femtoseconds elapse in each second than there have been hours since the big bang.)

A laser puts out a continuous beam of only one color, but pulse that laser, and you get a mixture of colors in each flash. The spectrum of a femtosecond pulse is a bizarre thing to see; millions of sharp lines spanning the rainbow, each line spaced exactly the same distance from its neighbors—like tick marks on a ruler. "That you could make a laser that pulses a billion times a second and whose constituent frequencies are all stable to one hertz is just short of unbelievable," Gibble said.

Diddams's group at NIST has built a rudimentary optical clockwork around mercury ions, which they immobilize in an electromagnetic trap. Because each atom is missing an electron, the ions carry a positive charge, so they repel one another, so collisions are no longer a problem. The device is stable to better than six parts in 10 to the 16th power over the course of a second. "Mercury is not an ideal element to use," Sullivan acknowledged. "The clock transition we use in it can shift with magnetic fields, which are hard to eliminate completely. But there is a transition in indium that looks attractive."

Groups at the Federal Institute of Physics and Metrology in Brunswick, Germany, and elsewhere are experimenting with uncharged calcium atoms. Because neutral atoms can be crammed more densely into the trap than ions can, the signal soars higher over the noise. And in 2015 a NIST-led team reported the successful demonstration of an optical lattice clock based on atoms of strontium 87; the uncertainty was pegged at two parts for 10 to the 18th power.

A German research group reported in 2016 on a precision almost as good, of 3.2 parts in 10 to the 18th power, from a system based on a single ion of ytterbium 171. The NIST group also used ytterbium for a lattice clock that achieved an accuracy of 1.6 parts in 10 to the 18th power

But there's that word again: accuracy. These new clocks "move away from the atomic definition of the second, which is based on the properties of cesium," Sullivan points out. For the newest and best clocks to be strictly accurate as keepers of the time to which we set our watches, that definition will have to change. Sullivan says the time committee of the International Bureau of Weights and Measures, which decides such things, accepted his proposal to allow "secondary" definitions that state the equivalence of a cesium frequency to that of other atoms. If the full assembly approves the idea, the definition of the second will be broadened but also weakened.

Clock builders will not get around relativity quite as easily. Clocks accurate to one part in 10 to the 17th power—which is about millisecond in three million years—will be easily thrown out of whack by two relativistic effects. First, there is time dilation: moving clocks run slow. "A frequency shift of 10 to the negative-17th power corresponds to a time dilation due to just walking speed," Gibble said.

The other confounder is gravity. The stronger its pull, the slower time passes. Clocks at the top of Mount Everest pull ahead of those at sea level by about 30 microseconds a year. When NIST researchers synchronized two quantum logic clocks and then raised one of them 33 centimeters, they found that the clocks were perceptibly out of sync. Raising a clock 10 centimeters changes its rate by one part in 10 to the 17th power. And elevation is relatively easy to measure, compared with variations in gravity caused by local geology, tides or even magma shifting around kilometers underground.

Ultimately, Gibble said, "If you take our ability to split spectral lines with microwave clocks and extrapolate to optical rulers, that puts you at uncertainties of order one in 10 to the 22nd power. I certainly would not claim that we are going to get there anytime soon, however." And there's no particular rush. No one has any idea how to transfer time that precisely between two clocks. And what good is a clock if you can't move it and can't check it against another?

8

TIMES OF OUR LIVES

This lesson looks at how humans experience and are affected by the passage of time. Neurologists and other researchers have begun to answer some of the most pressing questions raised by human experience in the fourth dimension. Examples of such questions include: Why does a watched pot never boil? Why does time fly when we're having fun? Why do people live longer than, for instance, hamsters?

Lesson 8 | Times of Our Lives

INTERVAL TIMING

⊿ The biopsychologist John Gibbon called time the "primordial context": a fact of life that has been felt by all organisms in every era. In human bodies, biological clocks keep track of seconds, minutes, days, months, and years.

⊿ When humans watch something that intrigues them, the time spent doing so seems to pass quickly, but when humans are bored, time drags. That's a quirk of a "stopwatch" in the brain—the so-called interval timer—that marks time spans of seconds to hours. As an example of this in action, the interval timer helps a baseball player figure out how fast he has to run to catch a baseball.

IMPRECISION

The precision of interval timers has been found to range from 5% to 60%. They don't work very well when a person is distracted or tense. Timing errors become worse as an interval gets longer. That's why we rely on cell phones and wristwatches to tell time.

- Interval timing enlists the higher cognitive powers of the cerebral cortex, the part of the brain that governs perception, memory, and conscious thought. When a driver approaches a yellow traffic light, for example, the driver might time how long it has been yellow and compare that with a memory of how long yellow lights usually last.

- After that, the driver "has to make a judgment about whether to put on the brakes or keep driving," according to the Cleveland Clinic's Stephen M. Rao. Rao's studies with functional magnetic resonance imaging (fMRI) have pointed to the parts of the brain engaged in each of those stages.

- Inside the fMRI machine, volunteers listen to two pairs of tones and decide whether the interval between the second pair is shorter or longer than the interval between the first pair. The brain structures that are involved in the task consume more oxygen than those that are not involved, and the fMRI scan records changes in blood flow and oxygenation once every 250 milliseconds.

- Rao says that "the very first structures that are activated are the basal ganglia." Long associated with movement, this collection of brain regions has become a prime suspect in the search for the interval-timing mechanism as well.

A **NETWORK** IN THE BRAIN

striatal neuron

- One area of the basal ganglia, the striatum, hosts a population of conspicuously well-connected nerve cells that receive signals from other parts of the brain. The long arms of these striatal cells are covered with between 10,000 and 30,000 spines, each of which gathers information from a different neuron in another location.

- If the brain acts like a network, then the striatal spiny neurons are critical nodes. "This is one of only a few places in the brain where you see thousands of neurons converge on a single neuron," according to Warren H. Meck of Duke University.

- Striatal spiny neurons are central to an interval-timing theory Meck developed with Gibbon. The theory posits a collection of neural oscillators in the cerebral cortex: nerve cells firing at different rates, without regard to their neighbors' tempos.

- In fact, many cortical cells are known to fire at rates between 10 and 40 cycles per second without external provocation. "All these neurons are oscillating on their own schedules," per Meck, "like people talking in a crowd. None of them are synchronized."

- The cortical oscillators connect to the striatum via millions of signal-carrying arms, so the striatal spiny neurons can eavesdrop on all those haphazard "conversations." Then something—like a yellow traffic light, for instance—gets the cortical cells' attention.

- The stimulation prompts all the neurons in the cortex to fire simultaneously, causing a characteristic spike in electrical output some 300 milliseconds later. This attentional spike acts like a starting gun, after which the cortical cells resume their disorderly oscillations.

- But because they've begun simultaneously, the cycles now make a distinct, reproducible pattern of nerve activation from moment to moment. The spiny neurons monitor those patterns, which help them to "count" elapsed time.

- At the end of a specified interval—when the traffic light turns red, for example—a part of the basal ganglia called the substantia nigra sends a burst of the neurotransmitter dopamine to the striatum. The dopamine burst induces the spiny neurons to record the pattern of cortical oscillations they receive at that instant, like a flashbulb exposing the interval's cortical signature on the spiny neurons' film.

▲ According to Meck, "There's a unique time stamp for every interval" imaginable. Once a spiny neuron has learned the time stamp of the interval for a given event, subsequent occurrences of the event prompt both the "firing" of the cortical starting gun and a burst of dopamine at the beginning of the interval. The dopamine burst now tells the spiny neurons to start tracking the patterns of cortical impulses that follow.

▲ When the spiny neurons recognize the time stamp marking the end of the interval, they send an electrical pulse from the striatum to another brain center, called the thalamus. The thalamus, in turn, communicates with the cortex, and the higher cognitive functions—such as memory and decision-making—take over. Hence, the timing mechanism loops from the cortex to the striatum to the thalamus and back to the cortex again.

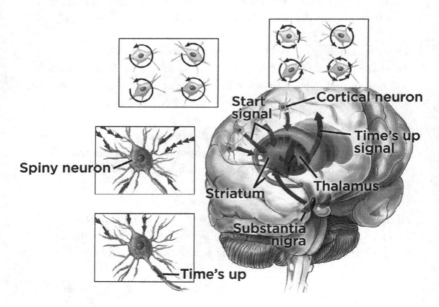

DISEASES, DRUGS, AND **HORMONES**

⊿ If dopamine bursts play an important role in framing a time interval, then diseases and drugs that affect dopamine levels should also disrupt that loop. That is what Meck and others have found.

⊿ Patients with untreated Parkinson's disease, for example, release less dopamine into the striatum, and their clocks run slow. Marijuana also lowers dopamine availability and slows time. Recreational stimulants such as cocaine and methamphetamine increase the availability of dopamine and make the interval clock speed up, so that time seems to expand.

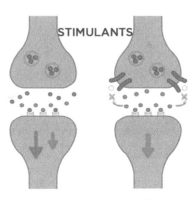

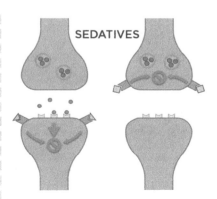

STIMULANTS

Nicotine ●
Mimics the action of the neurotransmitter acetylcholine

Cocaine/ecstasy ○
Increases release and blocks reuptake of dopamine and 5-HT

SEDATIVES

Benzodiazepines △
Binds to GABA receptors, causing hyperpolarization (postsynaptic)

THC/cannabis ▫
Binds cannabinoid receptors, causing hyperpolarization (presynaptic)

▲ Adrenaline and other stress hormones make the clock speed up, too, which may be why a second can feel like an hour during unpleasant situations. States of deep concentration or extreme emotion may flood the system or bypass it altogether; in such cases, time may seem to stand still or not to exist at all.

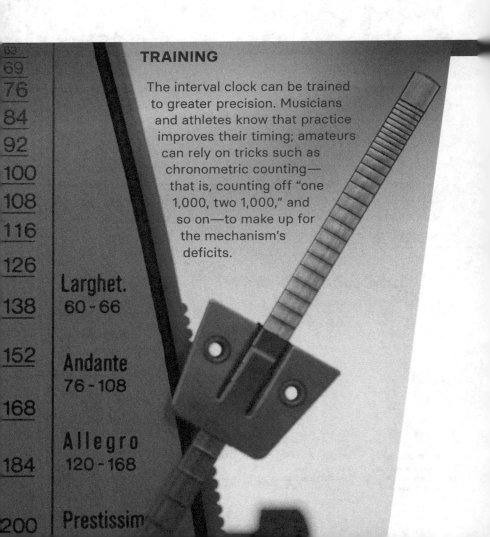

TRAINING

The interval clock can be trained to greater precision. Musicians and athletes know that practice improves their timing; amateurs can rely on tricks such as chronometric counting—that is, counting off "one 1,000, two 1,000," and so on—to make up for the mechanism's deficits.

Larghet. 60 - 66

Andante 76 - 108

Allegro 120 - 168

Prestissim

THE **CIRCADIAN** CLOCK

- A more rigorous timepiece chimes in at intervals of 24 hours. The circadian clock tunes our bodies to the cycles of sunlight and darkness that are caused by the Earth's rotation. It helps to program the daily habit of sleeping at night and waking in the morning. Its influence extends much further, however.

- Body temperature regularly peaks in the late afternoon or early evening and bottoms out a few hours before we rise in the morning. Blood pressure typically starts to surge between 6:00 and 7:00 am. Secretion of the stress hormone cortisol is 10 to 20 times higher in the morning than at night. Urination and bowel movements are generally suppressed at night and then pick up again in the morning.

- The circadian timepiece is more like a clock than a stopwatch because it runs without the need for a stimulus from the external environment. Studies of volunteers who live in caves for prolonged studies and other human guinea pigs have demonstrated that circadian patterns persist even in the absence of daylight, occupational demands, and caffeine.

- Moreover, they're expressed in every cell of the body. Confined to a petri dish under constant lighting, human cells still follow 24-hour cycles of gene activity, hormone secretion, and energy production. The cycles are hardwired, and they vary by as little as 1%—just minutes a day.

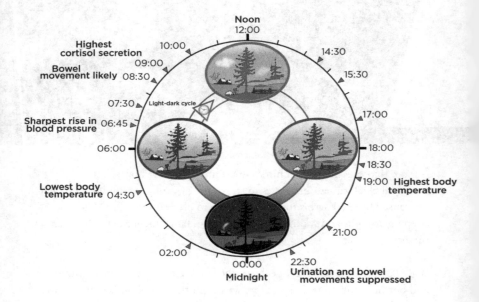

- Light isn't required to establish a circadian cycle, but it is needed to synchronize the phase of the hardwired clock with natural day and night cycles. Like an ordinary clock that runs a few minutes slow or fast each day, the circadian clock needs to be continually reset to stay accurate.

- Jet lag and shift work are exceptional conditions in which the innate circadian clock is abruptly thrown out of phase with the light-dark cycles or sleep-wake cycles. But the same thing can happen every year, albeit less abruptly, when the seasons change.

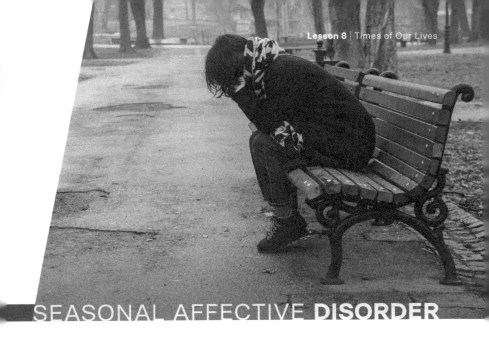

SEASONAL AFFECTIVE **DISORDER**

- Research shows that although bedtimes may vary, people tend to get up at about the same time in the morning year-round—usually because their dogs, kids, parents, or careers demand it. In the winter, at northern latitudes, that means many people wake up two to three hours before the sun makes an appearance. Their sleep-wake cycle is several time zones away from the cues they get from daylight.

- The mismatch between day length and daily life could explain the syndrome known as seasonal affective disorder, or SAD. In the US, SAD afflicts as many as one in 20 adults with depressive symptoms such as weight gain, apathy, and fatigue between October and March. The condition is 10 times more common in the north than the south.

- Although SAD occurs seasonally, some experts suspect it is actually a circadian problem. The research of Alfred J. Lewy suggests that SAD patients would come out of their depression if they could get up at the natural dawn in the winter. In his view, SAD isn't so much a pathology as evidence of an adaptive, seasonal rhythm in sleep-wake cycles.

BIOLOGY AND **MORTALITY**

- This lesson now turns to look at aging and mortality. People tend to relate aging to the diseases of aging—cancer, heart disease, osteoporosis, arthritis, and Alzheimer's, to name a few—as if the absence of disease would be enough to confer immortality. Biology suggests otherwise.

- Modern humans in wealthy countries have a life expectancy of more than 70 years. The life expectancy of an average mayfly, in contrast, is a day. Biologists are just beginning to explore why different species have different life expectancies.

- Comparisons within and among animal species, along with research on aging, have challenged many common assumptions about the factors that determine natural life span. The answer cannot lie solely with a species' genes: Worker honeybees, for example, last a few months, whereas queen bees live for years.

- However, genetics are important. This is exemplified by the fact that single-gene mutation in mice can produce a strain that lives up to 50% longer than usual.

- High metabolic rates can shorten life span, yet many species of birds, which have fast metabolisms, live longer than mammals of comparable body size. And big, slow-metabolizing animals don't necessarily outlast the small ones.

- Scientists in search of the limits to human life span have traditionally approached the subject from the cellular level rather than considering whole organisms. So far, the closest thing they have to a terminal timepiece is the so-called mitotic clock. The clock keeps track of cell division, or mitosis, the process by which a single cell splits into two.

- There seems to be a ceiling on how many times normal cells of the human body can divide. In culture, they will undergo 60 to 100 mitotic divisions. Then, they cease to grow, according to John Sedivy of Brown University, who explains, "They respire, they metabolize, they move, but they will never divide again."

- Cultured cells usually reach this state of senescence in a few months. Fortunately, most cells in the body divide much, much more slowly than cultured cells. Eventually—perhaps after 70 years or so—they, too, can end. Per Sedivy, the cells are counting "the number of cell divisions" rather than chronological time.

- Sedivy has shown that he could squeeze 20 to 30 more cycles out of human fibroblasts by mutating a single gene. This gene encodes a protein called p21, which responds to changes in structures called telomeres that cap the ends of chromosomes.

TELOMERES AND AGING

- Telomeres are made of the same stuff that genes are: DNA. They consist of thousands of repetitions of a six-base DNA sequence that doesn't code for any known protein. Each time a cell divides, chunks of its telomeres are lost.

- Young human embryos have telomeres between 18,000 and 20,000 bases long. By the time senescence kicks in, the telomeres are only 6,000 to 8,000 bases long.

- Biologists suspect that cells become senescent when telomeres shrink below some specific length. Titia de Lange has proposed an explanation for this link. In healthy cells, she showed, the chromosome ends are looped back on themselves like a hand tucked in a pocket.

- The "hand" is the last 100 to 200 bases of the telomere, which are single-stranded, not paired like the rest. With the help of more than a dozen specialized proteins, the single-stranded end is inserted into the double strands upstream for protection.

- If telomeres are allowed to shrink enough, "they can no longer do this looping trick," de Lange says. Untucked, a single-stranded telomere end is vulnerable to fusion with other single-stranded ends. The fusion wreaks havoc in a cell by stringing together all the chromosomes. That could be why Sedivy's mutated p21 cells died after they got in their extra rounds of mitosis.

CONCLUSION

- For now, the link between shortened telomeres and aging is tenuous at best. Experts point out that telomere length varies so much among individuals that it can't be used as a reliable indicator of biological age.

- In any case, most cells don't need to keep dividing to do their job—although white blood cells that fight infection are an exception. Many older people do die of infections that a younger body could withstand.

- In any case, telomere loss is just one of the numerous insults cells sustain when they divide, says Judith Campisi, a professor who has studied aging in-depth. For example, DNA is often damaged when it is replicated during cell division, so cells that have split many times are more likely to harbor genetic errors than young cells.

- "Cell division is very risky business," Campisi observes. Therefore, perhaps it's not surprising that the body puts a cap on mitosis. And cheating cell senescence probably wouldn't grant immortality.

> **ABOUT THIS LESSON**
>
> This lesson was adapted from the article "Times of Our Lives" by Karen Wright, an award-winning writer and editor.

Lesson 8 Transcript

TIMES OF OUR LIVES

This lesson was adapted from an article by Karen Wright, an award-winning writer and editor based near Seattle.

The biopsychologist John Gibbon called time the "primordial context," a fact of life that's been felt by all organisms in every era. For the morning glory that spreads its petals at dawn, for geese flying south in the autumn, for locusts swarming every 17 years and even for lowly slime molds sporing in daily cycles, timing is everything.

In human bodies, biological clocks keep track of seconds, minutes, days, months and years. They govern the split-second moves of a tennis serve and account for the hassle of jet lag and bouts of wintertime blues. Cellular chronometers may even decide when your time is up. Life ticks, then you die.

The pacemakers involved are as different as stopwatches and sundials. Some are accurate and inflexible, others less reliable but subject to conscious control. Some are set by planetary cycles, others by molecular ones. They're essential to the most sophisticated tasks the brain and body perform and timing mechanisms offer insights into aging and disease. Cancer, Parkinson's disease, seasonal depression and attention-deficit disorder have all been linked to disturbances in our biological clocks.

The physiology of these timepieces isn't completely understood but neurologists and other clock researchers have begun to answer some of the most pressing questions raised by human experience in the fourth dimension. Why, for example, does a watched pot never boil? Why does time fly when you're having fun? Why can all-nighters can give you indigestion? Or why do people live longer than hamsters? It's only a matter of time before clock studies resolve even more profound questions of temporal existence.

Lesson 8 Transcript | Times of Our Lives

You'll probably notice that if this lesson intrigues you, the time you spend listening to it will pass quickly. But it'll drag if you get bored. That's a quirk of a *stopwatch* in the brain—the so-called interval timer—that marks time spans of seconds to hours. The interval timer helps you figure out how fast you have to run to catch a baseball; it tells you when to clap to your favorite song; and it lets you sense how long you can lounge in bed after the alarm goes off.

Interval timing enlists the higher cognitive powers of the cerebral cortex, the part of the brain that governs perception, memory and conscious thought. When you approach a yellow traffic light while driving your car, for example, you time how long it's been yellow and compare that with a memory of how long yellow lights usually last. "Then you have to make a judgment about whether to put on the brakes or keep driving," says Stephen M. Rao, at the Cleveland Clinic. Rao's studies with functional magnetic resonance imaging (also called fMRI) have pointed to the parts of the brain engaged in each of those stages.

Inside the fMRI machine, volunteers listen to two pairs of tones and decide whether the interval between the second pair is shorter or longer than the interval between the first pair. The brain structures that are involved in this task consume more oxygen than those that are not involved and the fMRI scan records changes in blood flow and oxygenation once every 250 milliseconds. "When we do this, the very first structures that are activated are the basal ganglia," Rao says. Long associated with movement, this collection of brain regions has become a prime suspect in the search for the interval-timing mechanism as well.

One area of the basal ganglia, called the striatum, hosts a population of conspicuously well-connected nerve cells that receive signals from other parts of the brain. The long arms of these striatal cells are covered with between 10,000 and 30,000 spines, each of which gathers information from a different neuron in another location. If the brain acts like a network, then the striatal spiny neurons (as they are called) are critical nodes. "This is one of only a few places in the brain where you see thousands of neurons converge on a single neuron," says Warren Meck of Duke University.

Striatal spiny neurons are central to an interval-timing theory Meck developed with Gibbon. The theory posits a collection of neural oscillators in the cerebral cortex: nerve cells firing at different rates, without regard to their neighbors' tempos. In fact, many cortical cells are known to fire at rates between 10 and 40 cycles per second without external provocation. "All these neurons are oscillating on their own schedules," Meck observes, "like people talking in a crowd. None of them are synchronized."

The cortical oscillators connect to the striatum via millions of signal-carrying arms, so the striatal spiny neurons can eavesdrop on all those haphazard "conversations." Then something—say a yellow traffic light—gets the cortical cells' attention. The stimulation prompts all the neurons in the cortex to fire simultaneously, causing a characteristic spike in electrical output about 300 milliseconds later. This attentional spike acts like a starting gun, after which the cortical cells resume their disorderly oscillations. But because they've begun simultaneously, the cycles now make a distinct, reproducible pattern of nerve activation from moment to moment. The spiny neurons monitor these patterns, which help them to count elapsed time.

At the end of a specified interval—when the traffic light turns red, for example—a part of the basal ganglia called the substantia nigra sends a burst of the neurotransmitter dopamine to the striatum. The dopamine burst induces the spiny neurons to record the pattern of cortical oscillations they receive, at that instant, like a flashbulb exposing the interval's cortical signature on the spiny neurons' film. "There's a unique time stamp for every interval you can imagine," Meck says.

Once a spiny neuron has learned the time stamp of the interval for a given event, subsequent occurrences of the event prompt both the firing of the cortical starting gun and a burst of dopamine at the beginning of the interval. The dopamine burst now tells the spiny neurons to start tracking the patterns of cortical impulses that follow.

When the spiny neurons recognize the time stamp marking the end of the interval, they send an electrical pulse from the striatum to another brain center, called the thalamus. The thalamus, in turn, communicates with the cortex, and the higher cognitive functions—such as memory and decision making—can take over. Hence, the timing mechanism loops from the cortex to the striatum to the thalamus and back to the cortex again.

If dopamine bursts play an important role in framing a time interval, then diseases and drugs that affect dopamine levels should also disrupt that loop. And that's just what Meck and others have found. Patients with untreated Parkinson's disease, for example, release less dopamine into the striatum, and their clocks run slow. In trials, these people consistently underestimate the duration of time intervals.

Marijuana also lowers dopamine availability and slows time. Stimulants such as cocaine and methamphetamine increase the availability of dopamine and make the interval clock speed up, so that time seems to expand. Adrenaline and other stress hormones make the clock speed up, too, which may be why a second can feel like an hour during unpleasant situations.

States of deep concentration or extreme emotion may flood the system or bypass it altogether; in such cases, time may seem to stand still or not to exist at all. Because an attentional spike initiates the timing process, Meck suggests people with attention-deficit hyperactivity disorder might also have problems gauging the true length of intervals.

The interval clock can also be trained to greater precision. Musicians and athletes know that practice improves their timing and amateurs can rely on tricks such as chronometric counting (or saying "one one-thousand, two-one-thousand") to make up for the mechanism's deficits. Rao asks the volunteers who participate in his fMRI studies not to count, because it could activate brain centers related to language as well as those related to timing. But counting works, he says, it works well enough to expose cheaters. "The effect is so dramatic that we can tell whether they're counting or timing based just on the accuracy of their responses."

One of the virtues of the interval-timing stopwatch is its flexibility. You can start and stop it at will or ignore it completely. It can work subliminally or submit to conscious control. But it won't win any prizes for accuracy. The precision of interval timers has been found to range from 5% to 60%. They don't work too well if you're distracted or tense and timing errors get worse as an interval gets longer. That's why we rely on cell phones and wristwatches to tell time.

Fortunately, a more rigorous timepiece chimes in at intervals of 24 hours. The circadian clock tunes our bodies to the cycles of sunlight and darkness that are caused by the earth's rotation. It helps to program the daily habit of sleeping at night and waking in the morning, but its influence extends much further. Body temperature regularly peaks in the late afternoon or early evening and bottoms out a few hours before we rise in the morning. Blood pressure typically starts to surge between 6:00 am and 7:00 am. Secretion of the stress hormone cortisol is 10 to 20 times higher in the morning than at night. Urination and bowel movements are generally suppressed at night and then pick up again in the morning.

The circadian timepiece is more like a clock than a stopwatch because it runs without the need for a stimulus from the external environment. Studies of volunteers who live in caves for prolonged periods of time and studies of other human guinea pigs have demonstrated that circadian patterns persist even in the absence of daylight, or occupational demands or caffeine. Moreover, they're expressed in every cell of the body. Confined to a petri dish under constant lighting, human cells still follow 24-hour cycles of gene activity, hormone secretion and energy production. The cycles are hardwired, and they vary by as little as 1% by just minutes a day.

Light is not required to establish a circadian cycle, but it is needed to synchronize the phase of the hardwired clock with natural day and night cycles. Like an ordinary clock that runs a few minutes slow or fast each day, the circadian clock needs to be continually reset to stay accurate.

Neurologists have made great progress in understanding how daylight sets the clock. Two clusters of 10,000 nerve cells in the hypothalamus of the brain have long been considered the clock's locus. Decades of animal studies have demonstrated that these centers, each called a suprachiasmatic nucleus (or SCN), drive daily fluctuations in blood pressure, body temperature, activity level and alertness. The SCN also tells the brain's pineal gland when to release melatonin, a hormone which promotes sleep in humans and is secreted only at night.

Researchers have seen that dedicated cells in the retina of the eye transmit information about light levels to the SCN. These cells—a subset of those known as ganglion cells—operate completely independently of the rods and cones that mediate vision, and they're far less responsive to sudden changes in light. That sluggishness befits a circadian system. It would be no good if, say, watching fireworks or going to a movie matinee tripped the mechanism.

Scientists had assumed that the SCN somehow coordinated all the individual cellular clocks in the body's organs and tissues. But in the mid-1990s, researchers discovered four critical genes that govern circadian cycles in flies, mice and humans. Surprisingly, these genes turned up not just in the SCN but everywhere else, too. "These clock genes are expressed throughout the whole body, in every tissue," says Joseph Takahashi at the University of Texas Southwestern Medical Center. "We didn't expect that."

More recently, researchers found that the expression of more than 1,000 genes in the heart and liver tissue of mice varied in regular 24-hour periods. But the genes that showed these circadian cycles differed in the two tissues, and their expression peaked in the heart at different hours than in the liver. "They're all over the map," says Michael Menaker of the University of Virginia. "Some are peaking at night, some in the morning and some in the daytime."

Menaker has shown that specific feeding schedules can shift the phase of the liver's circadian clock, overriding the light-dark rhythm followed by the SCN. When lab rats that usually ate at will were fed just once a day, for example, peak expression of a clock gene in the liver shifted by 12 hours, whereas the same clock gene in the SCN stayed locked in sync with light schedules.

So, it makes sense that daily rhythms in feeding would affect the liver, given its role in digestion. Researchers think circadian clocks in other organs and tissues may respond to other external cues—including stress, exercise, maybe temperature changes—that occur regularly every 24 hours. The SCN still has authority over body temperature, blood pressure and other core rhythms is still secure. But this brain center is no longer thought to rule all the peripheral clocks throughout the body. "We have oscillators in our organs that can function independently of our oscillators in our brain," Takahashi says.

The autonomy of the peripheral clocks makes a phenomenon such as jet lag far more comprehensible. Whereas the interval timer, like a stopwatch, can be reset in an instant, circadian rhythms take days and sometimes weeks to adjust to a sudden shift in day length or time zone. A new schedule of light will slowly reset the SCN clock. But the other clocks may not follow its lead. The body is not only lagging, it's lagging at a dozen different places.

Jet lag doesn't last, presumably because all those different drummers are able to eventually sync up again. But shift workers, party animals, college students and other night owls face a worse chronological dilemma. They may be leading a kind of physiological double life. Even if they nap by day, their core rhythms are still ruled by the SCN—so the core functions continue "sleeping" at night. "You can will your sleep cycle earlier or later," says Alfred J. Lewy of the Oregon Health & Science University. "But you can't will your melatonin levels earlier or later, or your cortisol levels, or your body temperature."

Meanwhile, their schedules for eating and exercising could be setting their peripheral clocks to entirely different phases from either the sleep-wake cycle or the light-dark cycle. With their body living in so many time zones at once, it's no wonder shift workers have an increased incidence of heart disease, gastrointestinal complaints and, of course, sleep disorders.

Lesson 8 Transcript | Times of Our Lives

Jet lag and shift work are exceptional conditions in which the innate circadian clock is abruptly thrown out of phase with the light-dark cycles or sleep-wake cycles. But the same thing can happen every year, albeit less abruptly, when the seasons change. Research shows that although bedtimes may vary, people tend to get up at about the same time in the morning year-round; usually because their dogs, kids, parents or careers demand it. In the winter, at northern latitudes, that means many people wake up two to three hours before the sun makes an appearance. Their sleep-wake cycle is several time zones away from the cues they get from daylight.

The mismatch between day length and daily life could explain the syndrome known as seasonal affective disorder, or SAD. In the US, SAD afflicts as many as one in 20 adults with depressive symptoms such as weight gain, apathy and fatigue between October and March. The condition is 10 times more common in the north than the south. Although SAD occurs seasonally, some experts suspect it is actually a circadian problem. Lewy's work suggests that SAD patients would come out of their depression if they could get up at the natural dawn in the winter. In his view, SAD isn't so much a pathology as evidence of an adaptive, seasonal rhythm in sleep-wake cycles. "If we adjusted our daily schedules according to the seasons, we might not have seasonal depression," Lewy says. "We got into trouble when we stopped going to bed at dusk and getting up at dawn."

If modern civilization doesn't honor seasonal rhythms, it's partly because human beings are among the least seasonally sensitive creatures around. SAD is nothing compared to the annual cycles other animals go through: hibernation, migration, molting and especially seasonal mating, the master metronome to which all other cycles keep time. It's possible that these seasonal cycles may also be regulated by the circadian clock, which is equipped to keep track of the length of days and nights. Darkness, as detected by the SCN and the pineal gland, prolongs melatonin signals in the long nights of winter and reduces them in the summer.

"Hamsters can tell the difference between a 12-hour day, when their gonads don't grow," Menaker says, "and a 12-hour-and-15-minute day, when their gonads do grow." If seasonal rhythms are so robust in other animals and if humans have the equipment to express them, then how did we ever lose them? "What makes you think we ever had them?" Menaker asks. "We evolved in the tropics."

Menaker's point is that many tropical animals do not exhibit dramatic patterns of annual behavior. They don't need them, because the seasons themselves vary so little. Most tropical animals' mate without regard to season because there's no "best time" to give birth. People, too, are always functionally in heat. As our ancestors gained greater control of their environment over the millennia, seasons probably became even less significant of an evolutionary force.

But one aspect of human fertility is cyclical: women and other female primates release eggs just once a month. The clock that regulates ovulation and menstruation is a well-documented chemical feedback loop that can be manipulated by hormone treatments such as birth-control. The reason for the specific duration of the menstrual cycle is unknown, though. The fact that it's the same length as the lunar cycle may be a coincidence. No convincing link has been found between the moon's radiant or gravitational energy and a woman's reproductive hormones. In that regard, the monthly menstrual clock remains a mystery—outdone perhaps only by the ultimate conundrum, mortality.

People tend to relate aging to the diseases of aging—such as cancer, heart disease, osteoporosis, arthritis and Alzheimer's, to name just a few—as if the absence of disease would be enough to confer immortality. Biology suggests otherwise. Modern humans in wealthy countries have a life expectancy of more than 70 years. The life expectancy of your average mayfly, in contrast, is one day. Biologists are just beginning to explore why different species have such different life expectancies. If days are numbered, what's doing the counting?

Comparisons within and among animal species, along with research on aging, have challenged many common assumptions about the factors that determine natural life span. The answer cannot lie solely with a species' genes. Worker honeybees, for example, last a few months, whereas queen bees live for years. Still, genetics are important: a single-gene mutation in mice can produce a strain that lives up to 50% longer than usual.

High metabolic rates can shorten life span, yet many species of birds, which have very fast metabolisms, live longer than mammals of comparable body size. And big, slow-metabolizing animals don't necessarily outlast the small ones. The life expectancy of a parrot is about the same as a humans. Among dog species, small breeds typically live longer than large ones.

Scientists in search of the limits to human life span have traditionally approached the subject from the cellular level rather than considering whole organisms. So far, the closest thing they have to a terminal timepiece is the so-called mitotic clock. The clock keeps track of cell division, or mitosis, the process by which a single cell splits into two. The mitotic clock is like an hourglass in which each grain of sand represents one episode of cell division. Just as there are a finite number of grains in an hourglass, there seems to be a ceiling on how many times normal cells of the human body can divide. In cell culture, they will undergo 60 to 100 mitotic divisions, then call it quits.

"All of a sudden they just stop growing," says John Sedivy of Brown University. "They respire, they metabolize, they move, but they will never divide again." Cultured cells usually reach this state of senescence in a few months. Fortunately, most cells in the body divide much, much more slowly than cultured ones. Eventually—perhaps after 70 years or so—they, too, can end.

"What the cells are counting is not chronological time," Sedivy says. "It's the number of cell divisions." Sedivy has shown that he could squeeze 20 to 30 more cycles out of human fibroblasts cells by mutating a single gene. This gene encodes a protein called p21, which responds to changes in structures called telomeres that cap the ends of chromosomes.

Telomeres are made of the same stuff that genes are— DNA. They consist of thousands of repetitions of a six-base DNA sequence that doesn't code for any known protein. Each time a cell divides, chunks of its telomeres are lost. Human embryos have telomeres between 18,000 and 20,000 bases long. By the time senescence kicks in, the telomeres are only 6,000 to 8,000 bases long.

Biologists suspect that cells become senescent when telomeres shrink below some specific length. Titia de Lange of the Rockefeller University has proposed an explanation for this link. In healthy cells, she showed, the chromosome ends are looped back on themselves like a hand tucked in a pocket. The "hand" is the last 100 to 200 bases of the telomere, which are single-stranded, not paired like the rest. With the help of more than a dozen specialized proteins, the single-stranded end is inserted into the double strands upstream for protection.

If telomeres are allowed to shrink enough, "they can no longer do this looping trick," de Lange says. Untucked, a single-stranded telomere end is vulnerable to fusion with other single-stranded ends and this fusion wreaks havoc in a cell by stringing together all the chromosomes. That could be why Sedivy's mutated p21 cells died after they got their extra rounds of mitosis.

Other cells bred to ignore short telomeres have turned cancerous. The job of normal p21 and telomeres themselves may be to stop cells from dividing so much that they die or become malignant. Cellular senescence could actually be prolonging human life rather than spelling its doom. It might be cells' imperfect defense against malignant growth and certain death.

"Our hope is that we'll gain enough information from this reductionist approach to help us understand what's going on in the whole person," de Lange comments. For now, the link between shortened telomeres and aging is tenuous at best, although you wouldn't know that from outsized claims by some companies claiming they can predict life span by measuring the length of a person's telomeres. Experts point out that telomere length varies so much among individuals that it can't be used as a reliable indicator of biological age. In any case, most cells don't need to keep dividing to do their job, although white blood cells that fight infection are an exception. And many older people do die of infections that a younger body could withstand.

"Senescence probably has nothing to do with the nervous system," Sedivy says, because most nerve cells do not divide. "On the other hand, it might very well have something to do with the aging of the immune system." In any case, telomere loss is just one of the numerous insults that cells sustain when they divide, says Judith Campisi, a professor at the Buck Institute for Research on Aging in California.

DNA often gets damaged when it is replicated during cell division, so cells that have split many times are more likely to harbor genetic errors than young cells. Genes related to aging in animals and people often code for proteins that prevent or repair these mistakes. And with each mitotic episode, the by-products of copying DNA build up in cell nuclei, complicating subsequent bouts of replication.

"Cell division is very risky business," Campisi says. So perhaps it's not surprising that the body puts a cap on mitosis. And cheating cell senescence probably wouldn't grant immortality. Once the grains of sand have fallen through the mitotic hourglass, there's no point in turning it over again.

9
REMEMBERING WHEN

In the course of evolution, humans have developed a biological clock set to the alternating rhythm of light and dark brought by day and night. This clock, located in the brain's hypothalamus, governs what neuroscientist Antonio Damasio calls "body time." However, Damasio says there's another kind of time altogether, which he calls "mind time." Mind time has to do with how we experience the passage of time and how we organize chronology, which are the subjects of this lesson.

SUBJECTIVITY OF TIME

- Despite the steady tick of the clock, durations can seem fast or slow, short or long. And this variability can happen on different scales, from decades, seasons, weeks and hours, down to the tiniest intervals of music—the span of a note or the moment of silence between two notes. We also place events in time, deciding when they occurred, in which order, and on what scale.

- How mind time relates to the biological clock of body time is unknown. It's also not clear whether mind time depends on a single timekeeping device or if our experiences of duration and temporal order rely primarily, or even exclusively, on how we process information.

- If the latter alternative is true, mind time must be determined by the attention we give to events and the emotions we feel when they occur. It must also be influenced by the manner in which we record those events and the inferences we make as we perceive and recall them.

TIME IN THE BRAIN

- Damasio says he was first drawn to the problems of time processing through his work with neurological patients. People who sustain damage to regions of the brain involved in learning and recalling new facts develop major disturbances in their ability to place past events in the correct era and sequence. Moreover, these people with amnesia lose the ability to estimate the passage of time accurately at the scale of hours, months, years, and decades.

- Their biological clock, on the other hand, often remains intact, and so can their abilities to sense brief durations lasting a minute or less and to order them properly. At the very least, the experiences of these patients suggest that the processing of time and certain types of memory must share some common neurological pathways.

- The association between amnesia and time can be seen most dramatically in cases of permanent brain damage to the hippocampus, a region of the brain important to memory, and to the nearby temporal lobe, through which the hippocampus maintains its two-way communication with the rest of the cerebral cortex. Damage to the hippocampus prevents the creation of new factual memories, and the ability to form memories is an indispensable part of the construction of a sense of our own chronology.

- The memories that the hippocampus helps to create are distributed across neural networks located in parts of the cerebral cortex related to the material being recorded: areas dedicated to visual impressions, sounds, tactile information, and so forth. These networks must be activated to both lay down and recall a memory; when they're destroyed, patients can't recover long-term memories, a condition known as retrograde amnesia.

- The memories most markedly lost in retrograde amnesia are precisely those that bear a time stamp: recollections of unique events that happened in a particular context on a particular occasion. For instance, the memory of one's wedding bears a time stamp.

- A different but related kind of recollection—for instance, that of the concept of marriage—carries no such date with it. The temporal lobe cortex that surrounds the hippocampus is critical for making and recalling such memories.

- In patients who sustain damage to the temporal lobe cortex, years and even decades of autobiographical memory can be expunged irrevocably. Viral encephalitis, stroke, and Alzheimer's disease are among the neurological insults responsible for the most profound impairments.

AUTOBIOGRAPHICAL TIME LINES

- How the brain assigns an event to a specific time and then puts that event in a chronological sequence—or fails to do so—is still a mystery. We know only that both the memory of facts and the memory of spatial and temporal relations between those facts are involved.

- Damasio and his colleagues Daniel Tranel and Robert Jones decided to investigate how an autobiographical time line is established. By looking at people with different kinds of memory impairment, they hoped to identify what parts of the brain are required to place memories in the correct epoch.

- They selected four groups of participants, 20 people in total. The groups were broken down as such:

  - Patients with amnesia caused by damage in the temporal lobe.

  - Patients with amnesia caused by damage in the basal forebrain, another area relevant for memory.

  - Patients without amnesia who had damage in places other than the temporal lobe or basal forebrain.

  - Control subjects without neurological disease; these individuals had normal memories and were matched to the patients in terms of age and level of education.

- Every participant completed a detailed questionnaire about key events in their life. The investigators asked them about parents, siblings, various other relatives, schools, friendships, and professional activities. The team then verified the answers with relatives and records. They also established what the participants remembered of key public events.

- Then the team had each participant place a customized card that described a specific personal or public event on a board that laid out a year-by-year and decade-by-decade time line for the 1900s. The setup permitted a measurement of the accuracy of time placement.

THE RESULTS

- Predictably, the amnesiac patients differed from the controls. People with normal memory were relatively accurate in their time placements: On average, they were wrong by just 1.9 years.

- Patients with amnesia from basal forebrain damage made far more errors. Although they recalled the event exactly, they were off the mark by an average of 5.2 years. Their recall of events was superior to that of people with amnesia who had temporal lobe damage. However, patients with temporal lobe damage were nonetheless more accurate with regard to time stamping: They were off by an average of only 2.9 years.

- The results suggest that time stamping and event recall are processes that can be separated. More intriguingly, the outcome indicates that the basal forebrain may be critical in helping to establish the context that allows us to place memories in the right era. This notion is in keeping with the clinical observation of basal forebrain patients.

ROPE

Alfred Hitchcock's 1948 film *Rope* was shot in continuous, unedited 10-minute takes, and it is an interesting example of how people perceive the passage of time. In an interview with François Truffaut in 1966, Hitchcock stated that the story begins at 7:30 pm and ends 105 minutes later at 9:15. Yet the film consists of eight reels of 10 minutes each, making for a total of 80 minutes.

Despite the missing 25 minutes, the film never seems shorter than it should. Factors for this may include the visibility of the New York City skyline as night falls and the emotional content of the material: When we're uncomfortable or worried, we often experience time more slowly.

Additionally, a dinner party occurs in the film. The actual time during which food is served is about two reels. But viewers attribute more time to that sequence because we know that neither the hosts nor the guests, who look cool, polite, and unhurried, could swallow down dinner at such speed.

Finally, there are no jump cuts within each 10-minute reel. To join each segment to the next, Hitchcock finished most takes with a close-up on an object. Each interruption may contribute to the elongation of time because we're used to interpreting breaks in the continuity of visual perception as a lapse in the continuity of time.

MENTAL TIME LAG

- Most people don't have to grapple with the large gaps of memory or chronological confusion that many of Antonio Damasio's patients do. However, we all share a strange mental time lag, a phenomenon first brought to light in the 1970s by the neurophysiologist Benjamin Libet of the University of California, San Francisco.

- In one experiment, Libet documented a gap between the time someone was conscious of the decision to flex his finger and the time his brain waves indicated that a flex was imminent. The brain activity occurred a third of a second before the person consciously decided to move his finger.

- Libet performed another experiment in patients undergoing brain surgery. He found that a mild electrical charge to the cortex produced a tingling in a patient's hand a full half a second after the stimulus was applied.

- Although experts disagree how to interpret the findings, it's apparent that a lag exists between the beginning of the neural events leading to awareness and the moment one actually experiences the consequence of those events.

- This finding may seem strange at first glance, and yet the reasons for the delay are fairly obvious. It takes time for the physical signals to modify the sensory detectors of an organ such as the retina. It takes time for the resulting electrochemical modifications to be transmitted as signals to the central nervous system.

- It also takes time to generate a neural pattern in the brain's sensory maps. Finally, it takes time to relate the neural map of the event and the mental image arising from it to the neural map and image of the self—that is, the notion of who we are. This is the last and critical step without which the event will never become conscious.

▲ This delay covers mere milliseconds, but it is present nonetheless, which raises the question: Why can't we perceive it? One possible explanation is that because we have similar brains and they work similarly, we're all hopelessly late for consciousness and no one notices it.

▲ But perhaps other reasons apply. At the microtemporal level, the brain manages to "antedate" some events so that delayed processes can appear less delayed and differently delayed processes can appear to have similar delays. This possibility, which Libet contemplated, may explain why we maintain the illusion of continuity of time and space when our eyes move quickly from one target to another.

▲ We notice neither the blur caused by the eye movement nor the time it takes to get the eyes from one place to the other. Patrick Haggard and John C. Rothwell suggest that the brain predates the perception of the target by as much as 120 milliseconds, thereby giving us the perception of seamless viewing.

▲ The brain's ability to edit visual experiences and to impart a sense of volition after neurons have already acted is an indication of its exquisite sensitivity to time. Although our understanding of mind time is incomplete, we're gradually coming to know more about why we experience time so variably and about what the brain needs to create a time line.

ABOUT THIS LESSON

This lesson was adapted from the article "Remembering When" by Antonio Damasio, director of the Brain and Creativity Institute at the University of Southern California. He is the author of numerous scientific articles and books, among them *Descartes' Error* and *Self Comes to Mind*.

Lesson 9 Transcript
REMEMBERING WHEN

This lesson was adapted from an article by Antonio Damasio, director of the Brain and Creativity Institute at the University of Southern California. He is the author of numerous scientific articles and books, among them *Descartes' Error* and *Self Comes to Mind*.

We wake up to time, courtesy of an alarm clock, and go through a day run by time—the meeting, the visitors, the conference call, the luncheon are all set to begin at a particular hour. We can coordinate our own activities with those of others because we all implicitly agree to follow a single system for measuring time, one based on the rise and fall of daylight.

In the course of evolution, humans have developed a biological clock set to this alternating rhythm of light and dark. The clock, located in the brain's hypothalamus, governs what neuroscientist Antonio Damasio calls "body time." But Damasio says there's another kind of time altogether, which he calls "mind time." Mind time has to do with how we experience the passage of time and how we organize chronology.

Despite the steady tick of the clock, duration can seem fast or slow, short or long. And this variability can happen on different scales, from decades, seasons, weeks and hours, down to the tiniest intervals in music—the span of a note or the moment of silence between two notes. We also place events in time, deciding when they occurred, in which order and on what scale, whether that of a lifetime or of a few seconds.

How mind time relates to the biological clock of body time is unknown. It's also not clear whether mind time depends on a single timekeeping device or if our experiences of duration and temporal order rely primarily, or even exclusively, on how we process information. If the latter alternative is true, mind time must be determined by the attention we give to events and the emotions we feel when they occur. It must also be influenced by the manner in which we record those events and the inferences we make as we perceive and recall them.

Damasio says he was first drawn to the problems of time processing through his work with neurological patients. People who sustain damage to regions of the brain involved in learning and recalling new facts develop major disturbances in their ability to place past events in the correct era and sequence. Moreover, people with amnesia lose the ability to estimate the passage of time accurately at the scale whether it is hours, months, years and decades.

Their biological clock, on the other hand, often remains intact, and so can their abilities to sense brief durations lasting a minute or less and to order them properly. At the very least, the experiences of these patients suggest that the processing of time and certain kinds of memory must share some common neurological pathways.

The association between amnesia and time can be seen most dramatically in cases of permanent brain damage to the hippocampus, a region of the brain important to memory, and to the nearby temporal lobe—that is the area where the hippocampus maintains its two-way communication with the rest of the cerebral cortex. Damage to the hippocampus prevents the creation of new factual memories.

The ability to form memories is an indispensable part of the construction of a sense of our own chronology. We build our time line event by event and we connect personal happenings to those that occur around us. When the hippocampus is impaired, patients become unable to hold factual memories for longer than about one minute or make lasting new memories. These patients have what's called anterograde amnesia.

Intriguingly, the memories that the hippocampus helps to create aren't actually stored within the hippocampus. Instead, they're distributed across neural networks located in parts of the cerebral cortex related to the material being recorded. So, whether it is areas dedicated to visual impressions, sounds, tactile information, and so forth, these networks must be activated to both lay down and recall a memory; when they're destroyed, patients can't recover a long-term memory and that condition known as retrograde amnesia.

The memories most markedly lost in retrograde amnesia are precisely those that bear a time stamp: recollections of unique events that happened in a particular context on a particular occasion. For instance, the memory of one's wedding bears a time stamp. A different but related kind of recollection—say, that of the concept of marriage—carries no particular date with it. The temporal lobe cortex that surrounds the hippocampus is critical for making and recalling such memories.

In patients who sustain damage to the temporal lobe cortex, years and even decades of autobiographical memory can be expunged irrevocably. Viral encephalitis, stroke and Alzheimer's disease are among the neurological insults responsible for the most profound impairments. For one such patient, whom Damasio and his colleagues studied for 25 years, the time gap went almost all the way to the cradle. When this patient was 46, he sustained damage to both the hippocampus and to parts of the temporal lobe. So, he had both anterograde and retrograde amnesia; that is, he couldn't form new factual memories and he also couldn't recall old ones. The patient inhabited a permanent present, unable to remember what happened a minute earlier or 20 years before.

Indeed, he had no sense of time at all. He couldn't tell his doctors the date, and when asked to guess, his responses were wild—as disparate as 1942 and 2013. He could guess time more accurately if he had access to a window and could approximate it based on light and shadows. But if he was deprived of a watch or a window, morning was no different from afternoon, and night was no different from day; the clock of body time was no help. This patient couldn't state his age, either. He could guess, but the guess tended to be wrong.

Two of the few specific things he knew for certain were that he was married and that he was the father of two children. But when did he get married? He couldn't say. When were the children born? He didn't know. He couldn't place himself in the time line of his family life. He was in fact married, but his wife had divorced him more than two decades before and his children had long been married and had children of their own.

How the brain assigns an event to a specific time and then puts that event in a chronological sequence—or in the case of Damasio's patient, fails to do so—is still a mystery. We know only that both the memory of facts and the memory of spatial and temporal relations between those facts are involved. Damasio and his colleagues Daniel Tranel and Robert Jones decided to investigate how an autobiographical time line is established. By looking at people with different kinds of memory impairment, they hoped to identify what parts of the brain are required to place memories in the correct epoch.

They selected four groups of participants: 20 people total. The first group consisted of patients with amnesia caused by damage to the temporal lobe. Patients with amnesia caused by damage in the basal forebrain, another area relevant for memory, made up the second set. The third group was composed of patients without amnesia who had damage in places other than the temporal lobe or basal forebrain. And finally, the control subjects were people without neurological disease who had normal memories and who were matched to the patients in terms of age and level of education.

Every participant completed a detailed questionnaire about key events in their life. The investigators asked them about their parents, siblings and other various relatives, their schools, friendships and professional activities. The team then verified the answers with relatives and other records. They also established what the participants remembered of key public events, such as the election of officials, wars, natural disasters, and prominent cultural developments.

Then the team had each participant place a customized card that described a specific personal or public event on a board that laid out a year-by-year and decade-by-decade time line for the 1900s. For the participants, the situation was an experience similar to playing the board game *Life*. For the investigators, the setup permitted a measurement of the accuracy of time placement.

Predictably, the patients with amnesia differed from the controls. People with normal memory were relatively accurate in their time placements: on average, they were wrong by just 1.9 years. Patients with amnesia from basal forebrain damage made far more errors. Although they recalled the event exactly, they were off the mark by an average of 5.2 years. Their recall of events was superior to that of people with amnesia from temporal lobe damage. But patients with temporal lobe damage were nonetheless more accurate with regard to time stamping—they were off by an average of only 2.9 years.

So, the results suggest that time stamping and recall of events are processes that can be separated. More intriguingly, the outcome indicates that the basal forebrain may be critical in helping to establish the context that allows us to place memories in the right era. This notion is in keeping with the clinical observation of basal forebrain patients. Unlike certain of their counterparts with temporal lobe damage, these patients do learn new facts. But they often recall the facts they've just learned in the incorrect order, reconstructing sequences of events in a fictional narrative that can change from occasion to occasion.

Now, most of us don't have to grapple with the large gaps of memory or the chronological confusion that many of Antonio Damasio's patients do. But we all do share a strange mental time lag, a phenomenon first brought to light in the 1970s by the neurophysiologist Benjamin Libet of the University of California, San Francisco. In one experiment, Libet documented a gap between the time someone was conscious of the decision to flex his finger (and recorded the exact moment of that awareness) and the time his brain waves indicated that a flex was imminent. The brain activity occurred a full third of a second before the person consciously decided to move his finger.

Libet performed another experiment in patients undergoing brain surgery. For some such operations, patients are kept awake so the surgeons can map the functional areas of the person's brain. By lightly probing parts of the primary sensory cortex, for instance, surgeons can tell which part of the brain receives signals from the hand or the foot. In Libet's experiment, he found that a mild electrical charge to the cortex produced a tingling in a patient's hand a full half a second after the stimulus was applied.

Although experts disagree how to interpret the findings, it's apparent that a lag exists between the beginning of the neural events leading to awareness and the moment one actually experiences the consequence of those events.

The finding may seem strange at first glance, and yet the reasons for the delay are fairly obvious. It takes time for the physical signals to modify the sensory detectors of an organ such as the retina. It takes time for the resulting electrochemical modifications to be transmitted as signals to the central nervous system. It takes time to generate a neural pattern in the brain's sensory maps. Finally, it takes time to relate the neural map of the event and the mental image arising from it to the neural map and image of the self—that is, the notion of who we are—the last and critical step without which the event will never become conscious.

So, we're talking about nothing more than mere milliseconds, but there's a delay, nonetheless. Why can't we perceive it? One possible explanation is that because we have similar brains and they all work similarly, we're all hopelessly late for consciousness and no one ever notices it. But perhaps other reasons apply. At the micro-temporal level, the brain manages to antedate some events so that delayed processes can appear less delayed and differently delayed processes can appear to have similar delays.

This possibility, which Libet contemplated, may explain why we maintain the illusion of continuity of time and space even when our eyes move quickly from one target to another. We notice neither the blur caused by the eye movement nor the time it takes to get the eyes from one place to another. Patrick Haggard and John C. Rothwell of University College London suggest that the brain predates the perception of the target by as much as 120 milliseconds, thereby giving us the perception of seamless viewing.

The brain's ability to edit visual experiences and to impart a sense of volition after neurons have already acted is an indication of its exquisite sensitivity to time. Although our understanding of mind time is incomplete, we're gradually coming to know more about why we experience time so variably and about what the brain needs to create a time line.

The elasticity of time is perhaps best appreciated when we're watching a performance, such as a play, concert or film. The actual duration of the performance and its mental duration are different things.

To illustrate the factors that contribute to this varied experience of time, consider Alfred Hitchcock's 1948 film *Rope*. This technically remarkable work was shot in continuous, unedited 10-minute takes. Few features have been produced in their entirety using this approach. Orson Welles in *Touch of Evil*, Robert Altman in *The Player* and Martin Scorsese in *GoodFellas* employed long, continuous shots but not as consistently as in Rope. (By the way, the innovation earned praise for the director, but filming was a nightmare for everyone concerned, and Hitchcock used the method again only in part of his next film, which was called *Under Capricorn*.)

Hitchcock invented this technique for a practical reason. He was attempting to re-create a story that had been told in a play performed in continuous time. But he was limited to the amount of film that could be loaded into the camera, which was roughly enough for 10 minutes of action. Now, let's consider how *Rope*'s real time plays in our mind. In an interview with François Truffaut in 1966, Hitchcock stated that the story begins at 7:30 pm and ends 105 minutes later at 9:15. Yet the film consists of eight reels of 10 minutes each: a total of 80 minutes, including the credits at the beginning and end.

So, where did the missing 25 minutes go? Do we experience the film as shorter than 105 minutes? Not really. The film never seems shorter than it should, and a viewer doesn't get a sense of haste or clipping. On the contrary, for many people the film seems longer than its projection time.

Several aspects may account for this alteration of perceived time. First, most of the action takes place in the living room of a penthouse in summer, and the skyline of New York City is visible through a panoramic window. At the beginning of the film the light suggests late afternoon; by the end night has set in. Our daily experience of fading daylight makes us perceive the real-time action as taking long enough to cover the several hours of the coming of night, when in fact, those changes in light are artificially accelerated by Hitchcock.

In the same way, the nature and context of the actions depicted elicit other automatic judgments about time duration. After the murder, which occurs at the beginning of the film's first reel, the story focuses on an elegant dinner party hosted by the two murderers and attended by the relatives and friends of the victim. The actual time during which food is served in the film is about two reels. But viewers attribute more time to that sequence because we know that neither the hosts nor the guests, who look cool, polite and unhurried, could swallow down dinner at such speed.

When the action later splits—some guests converse in the living room in front of the camera, while others repair to the dining room to look at rare books—we sensibly attribute a longer duration to this offscreen episode than the few minutes it takes up in the actual film.

Another factor may also contribute to the deceleration of time. There are no jump cuts within each 10-minute reel; the camera glides slowly toward and away from each character. Yet to join each segment to the next, Hitchcock finished most takes with a close-up on an object. In most instances, the camera moves to the back of an actor wearing a dark suit, and the screen goes black for a few seconds; the next take begins as the camera pulls back away from the actor's back.

Although the interruption is brief and is not meant to signal a time break, it may nonetheless contribute to the elongation of time because we're used to interpreting breaks in the continuity of visual perception as a lapse in the continuity of time. Film-editing devices such as the dissolve and the fade often cause spectators to infer that time has passed between the preceding shot and the following one. In *Rope*, each of the seven breaks delays time by just a fraction of a second. But cumulatively for some viewers, the breaks may suggest that more time has passed.

The emotional content of the material may also extend time. When we are uncomfortable or worried, we often experience time more slowly because we focus on negative images associated with our anxiety.

Studies in Antonio Damasio's laboratory suggest that the brain generates images at faster rates when we're experiencing positive emotions (and perhaps this is why time flies when we're having fun) and the brain reduces the rate of image making during negative emotions. On a flight with heavy turbulence, for instance, you might experience the passage of time as achingly slow because your attention is directed to the discomfort and anxiety of the experience. The unpleasantness of the situation in *Rope* may similarly conspire to stretch time.

Rope provides a noticeable discrepancy between real time and the audience's perception of time. In so doing, it illustrates how the experience of duration is a construct. It's based on factors as varied as the way in which images are presented to us, as well as the conscious and unconscious inferences about the events, the content of the events being perceived, and even the emotional reaction those events provoke.

10
INCONSTANT CONSTANTS

Constants are the topic of this lesson. Such quantities as the velocity of light, c, Newton's constant of gravitation, G, and the mass of the electron, m_e, are assumed to be the same at all places and times in the universe. They form the framework around which the theories of physics are erected.

Lesson 10 | Inconstant Constants

TRYING TO EXPLAIN **CONSTANTS**

- It is unclear why constants take the special numerical values that they do. According to the International System of Units, c is 299,792,458; G is 6.673×10^{-11}; and m_e is $9.10938188 \times 10^{-31}$. These are numbers that follow no discernible pattern. The only thread running through the values is that if many of them were even slightly different, complex atomic structures such as living beings would be impossible.

- The desire to explain the constants has been one of the driving forces behind efforts to develop a complete unified description of nature, or "theory of everything." Physicists have hoped such a theory would show that each constant can have only one logically possible value. It would give an underlying order to the seeming arbitrariness of nature.

- In recent years, however, the status of the constants has grown more muddied, not less. Researchers have found that the best candidate for a theory of everything, the variant of string theory called M-theory, is self-consistent only if the universe has more than four dimensions of space and time—as many as seven more.

- One implication is that the constants we observe may not, in fact, be the truly fundamental ones. Those exist in the full higher-dimensional space, and we see only their three-dimensional "shadows."

- Meanwhile, physicists have also come to appreciate that the values of many of the constants may merely be the result of happenstance, acquired during random events and elementary particle processes early in the history of the universe. In fact, string theory allows for a vast number—10^{500}—of possible "worlds" with different self-consistent sets of laws and constants. Thus far, researchers have no idea why our combination exists.

- Continued study may reduce the number of logically possible worlds to just one, but we have to remain open to the unnerving possibility that our known universe is but one of many—a part of a multiverse. Different parts of the multiverse would exhibit different solutions to the theory, making our observed laws of nature just one of many systems of local bylaws.

INCONSTANCY

- The word *constant* itself may be a misnomer. Our constants could vary both in time and in space. If the extra dimensions of space were to change in size, the constants in our three-dimensional world would change with them. If we looked far enough out in space, we might begin to see regions where the so-called constants have settled into different values.

- Ever since the 1930s, researchers have speculated that the constants may not be constant. String theory gives this idea a theoretical plausibility and makes it all the more important for observers to search for deviations from constancy.

- Such experiments are challenging. The first problem is that the laboratory apparatus itself may be sensitive to changes in the constants. The size of all atoms could be increasing, but if the ruler being used to measure them is getting longer, too, it will be impossible to tell.

- Experimenters routinely assume their reference standards—rulers, masses, clocks—are fixed, but they can't do so when testing the constants. They must focus on constants that have no units, so their values are the same irrespective of the units system. An example is the ratio of two masses, such as the proton mass to the electron mass.

Lesson 10 | Inconstant Constants

- The second experimental problem is that measuring changes in the constants requires high-precision equipment that remains stable long enough to register any changes. Even atomic clocks could detect drifts in the fine-structure constant only over days or, at most, years.

- If α changed by more than four parts in 10^{15} over a three-year period, the best clocks would see it. None have. That may sound like an impressive confirmation of constancy, but three years is a cosmic eyeblink. Slow but substantial changes during the long history of the universe would've gone unnoticed.

- Fortunately, physicists have found other tests. During the 1970s, scientists at the French atomic energy commission noticed something peculiar about the isotopic composition of ore from a uranium mine in Oklo, Gabon: It looked like the waste products of a nuclear reactor. About 2 billion years ago, Oklo must have been the site of a natural reactor.

- In 1976, Alexander Shlyakhter of the Petersburg Nuclear Physics Institute in Russia noticed that the ability of a natural reactor to function depends crucially on the precise energy of a particular state of the samarium nucleus that facilitates the capture of neutrons. That energy depends sensitively on the value of α.

- Therefore, if the fine-structure constant had been slightly different, no chain reaction could have occurred. One did occur, though, which implies the constant hasn't changed by more than one part in 10^8 over the past 2 billion years.

- In 1962, James Peebles and Robert Dicke of Princeton University first applied similar principles to meteorites: The abundance ratios arising from the radioactive decay of different isotopes in these ancient rocks depend on α. The most sensitive constraint involves the beta decay of rhenium into osmium.

▲ Researchers found that, at the time the rocks formed, α was within two parts in 10^6 of its current value. This result is less precise than the Oklo data but goes back further in time, to the origin of our solar system 4.6 billion years ago.

LOOKING TO THE STARS

▲ To probe possible changes over even longer time spans, researchers must look to the stars. Light takes billions of years to reach our telescopes from distant astronomical sources. It carries a snapshot of the laws and constants of physics at the time when it started its journey or encountered material along the way.

▲ Astronomy first entered the constants story soon after the discovery of quasars in 1965. The idea was simple. Quasars were identified as bright sources of light located at huge distances from Earth. Because the path of light from a quasar to us is so long, it inevitably intersects the gaseous outskirts of young galaxies. That gas absorbs the quasar light at particular frequencies, imprinting a bar code of narrow lines onto the quasar spectrum.

- Whenever gas absorbs light, electrons within the atoms jump from a low-energy state to a higher one. These energy levels are determined by how tightly the atomic nucleus holds the electrons, which depends on the strength of the electromagnetic force between them—and therefore on the fine-structure constant alpha.

- If the constant was different at the time when the light was absorbed or in the particular region of the universe where it happened, then the energy required to lift the electrons would differ from that required today in lab experiments, and the wavelengths of the transitions seen in the spectra would differ. The way the wavelengths change depends critically on the orbital configuration of the electrons.

- For a given change in α, some wavelengths shrink, whereas others increase. The complex pattern of effects is hard to mimic by data-calibration errors, which makes the test astonishingly powerful.

- John Barrow and John Webb, who wrote the article this lesson is based on, needed some high-precision lab measurements to compare against the quasar spectra. Initial measurements were done by Anne Thorne and Juliet Pickering of Imperial College London, followed by groups at Lund Observatory in Sweden and the National Institute of Standards and Technology.

- A challenge was that previous observers had used so-called alkali-doublet absorption lines—pairs of absorption lines arising from the same gas, such as carbon or silicon. They compared the spacing between these lines in quasar spectra with lab measurements.

- However, this method failed to take advantage of one particular phenomenon: A change in α shifts not just the spacing of atomic energy levels relative to the lowest energy level, or ground state, but also the position of the ground state itself. Consequently, the highest precision observers achieved was only about one part in 10^4.

- In 1999, John Webb and Victor Flambaum of the University of New South Wales in Sydney came up with a method to take both effects into account and achieved 10 times higher sensitivity. Moreover, the method allows different species (for instance, magnesium and iron) to be compared, which allows additional cross-checks. Combined with modern telescopes and detectors, the new approach, known as the many-multiplet method, has enabled Webb and Barrow to test the constancy of α with unprecedented precision.

- When embarking on this project, they anticipated establishing that the value of the fine-structure constant long ago was the same as it is today; their contribution would simply be higher precision. To their surprise, the first results showed small but statistically significant differences.

- Further data confirmed this finding. Based on a total of 128 quasar absorption lines, they found an average increase in α of close to six parts in 1 million over the past 6 to 12 billion years.

NEW DATA

- By 2010, Barrow and Webb completed the analysis of a large amount of new data from the Very Large Telescope, or VLT, operated by the European Southern Observatory. They obtained 153 new measurements.

- All of the data their group had previously analyzed had come from the Keck telescopes on Mauna Kea in Hawaii. For these new VLT data, everything was different: the telescopes, the spectrograph, the detectors, and the software used for the initial stages of the data analysis. These VLT data therefore provided a beautiful cross-check with their results from the Keck telescopes.

- They thought it was possible that the new data would show no change in α at all or that they would show the same effect the Keck data did—with α appearing smaller at higher redshifts. However, the new VLT data showed not a smaller value of α at high redshift but a larger value. It was larger by just about the same amount as the Keck data are smaller.

- The researchers initially suspected they were seeing evidence for systematic problems in both data sets. Add the Keck and VLT samples together, and to a good approximation, the combined sample shows no change in α with redshift. The constants are really constant after all.

- However, if that's the explanation, it requires two different systematic effects, one for each telescope, such that both effects are, independently, of the same magnitude but opposite sign. This isn't impossible, but it's pretty unlikely.

- They did discover another curiosity, though. The Keck data cover a somewhat large portion of the sky in the Northern Hemisphere. It is large enough to ask: Could it be that α changes not with redshift but with position on the sky?

- A simple analysis suggested that might be the case. The VLT is in Chile and, on average, points to a very different part of the universe than the Keck telescopes do. It is possible that that is another coincidence, but that would make for two coincidences.

- When the old Keck and the new VLT samples are merged, the result is positively intriguing: the directional dependence becomes highly significant. Deriving such a result by chance appears to be extremely unlikely. If the result is a fluke, a subset of the data could be generating a rogue result.

- Despite extensive attempts, however, they've yet to find a combination of systematic effects in the data that could mimic a spatial dependence. Alpha appears to change spatially—across, perhaps, the entire observable universe.

- Any change with time is smaller and is currently below their detection sensitivity. In a study in *Science* in 2020, Webb and Barrow and their colleagues used four different measurements of the fine-structure constant going back 13 billion years and concluded that the spatial variation in alpha is real.

CONSEQUENCES

- The consequences of a variable constant are enormous, though only partially explored. Barrow and Webb's theory suggests that varying the fine-structure constant makes objects fall differently.

- Galileo predicted that bodies in a vacuum fall at the same rate no matter what they're made of—an idea known as the weak equivalence principle. But if α varies, that principle no longer holds exactly. The variations generate a force on all charged particles.

- The more protons an atom has in its nucleus, the more strongly it'll feel this force. If Webb and Barrow's quasar observations are correct, then the accelerations of different materials differ by about one part in 10^{14}, which is too small to measure currently.

- Going forward, the main scientific focus is on α, over the other constants of nature, simply because it's possible to build up a statistical sample of measurements, mapping the laws of physics throughout the distant cosmos in greater detail. If α is susceptible to change, however, other constants should vary as well, making the inner workings of nature fickler than scientists ever suspected.

ABOUT THIS LESSON

This lesson was adapted from the article "Inconstant Constants" by John Barrow, a professor of mathematical sciences at the University of Cambridge, and John Webb, a professor at the University of New South Wales.

Lesson 10 Transcript
INCONSTANT CONSTANTS

This lesson was adapted from an article by John Barrow, professor of mathematical sciences at the University of Cambridge, and John Webb, a professor at the University of New South Wales in Sydney, Australia.

Some things never change. Physicists call them the constants of nature. Such quantities as the velocity of light, which is known as c, Newton's constant of gravitation, known as G, and the mass of the electron, or m sub-e, are assumed to be the same at all places and all times in the universe. They form the framework around which the theories of physics are erected and they define the fabric of our universe. Physics has progressed by making ever more accurate measurements of their values.

And yet, remarkably, no one has ever successfully predicted or explained any of the constants. Physicists have no idea why constants take the special numerical values they do. According to the International System of Units, c is 299,792,458; G is 6.673 times 10 to the negative-11th power; and m sub-e is 9.10938188 times 10 to the negative-31st power. These numbers follow no discernible pattern. The only thread running through the values is that if many of them were even slightly different, complex atomic structures such as living beings would be impossible.

The desire to explain the constants has been one of the driving forces behind efforts to develop a complete unified description of nature, or a "theory of everything." Physicists have hoped such a theory would show that each constant can have only one logically possible value. It would give an underlying order to the seeming arbitrariness of nature.

In recent years, however, the status of the constants has grown more muddied, not less. Researchers have found that the best candidate for a theory of everything, the variant of string theory called M-theory, is self-consistent only if the universe has more than four dimensions of space and time—as many as seven more dimensions. One implication is that the constants we observe may not, in fact, be the truly fundamental ones. Those may exist in the full higher-dimensional space, and we see only their three-dimensional shadows.

Meanwhile, physicists have also come to appreciate that the values of many of the constants may merely be the result of happenstance, acquired during random events and elementary particle processes early in the history of the universe. In fact, string theory allows for a vast number—10 to the 500th power—of possible worlds with different self-consistent sets of laws and constants. Thus far researchers have no idea why our combination exists.

Continued study may reduce the number of logically possible worlds to just one, but we have to remain open to the unnerving possibility that our known universe is but one of many—a part of a multiverse. Different parts of the multiverse would exhibit different solutions to the theory, making our observed laws of nature just one of many systems of local bylaws.

No further explanation would then be possible for many of our numerical constants other than that they constitute a rare combination that permits consciousness to evolve. Our observable universe could be one of many isolated oases surrounded by an infinity of lifeless space—a surreal place where different forces of nature hold sway and particles such as electrons or structures such as carbon atoms and DNA molecules could be impossibilities. If you tried to venture into that outside world, you would cease to be.

Thus, string theory gives with the right hand and takes with the left. It was devised in part to explain the seemingly arbitrary values of the physical constants, yet the basic equations of the theory contain few arbitrary parameters. Yet so far string theory offers no explanation for the observed values of the constants.

Indeed, the word *constant* itself may be a misnomer. Our constants could vary both in time and in space. If the extra dimensions of space were to change in size, the constants in a three-dimensional world would change with them. If we looked far enough out in space, we might begin to see regions where the constants have settled into different values. Ever since the 1930s, researchers have speculated that the constants may not be constant. String theory gives this idea a theoretical plausibility and makes it all the more important for observers to search for deviations from constancy.

Such experiments are challenging. The first problem is that the laboratory apparatus itself may be sensitive to changes in the constants. The size of all atoms could be increasing, but if the ruler you're using to measure them is also getting longer, you'd never be able to tell.

Experimenters routinely assume their reference standards—rulers, masses, clocks—are fixed, but they can't do so when testing the constants. They must focus on constants that have no units, so their values are the same irrespective of the unit's system. An example is the ratio of two masses, such as the proton mass to the electron mass.

One ratio of particular interest combines the velocity of light, the electric charge on a single electron, Planck's constant, and the so-called vacuum permittivity. This quantity, called the fine-structure constant and represented with the Greek letter *alpha*, was first introduced in 1916 by Arnold Sommerfeld, a pioneer in applying the theory of quantum mechanics to electromagnetism. It quantifies the relativistic and quantum qualities of electromagnetic interactions involving charged particles in empty space. Measured to be equal to 1 over 137.03599976, or approximately 1 divided by 137. *Alpha* has endowed the number 137 with a legendary status among physicists (it is often part of the password that opens their cell phone lock screens).

If *alpha* had a different value, all sorts of vital features of the world around us would change. If the value were lower, the density of solid atomic matter would fall, molecular bonds would break at lower temperatures, and the number of stable elements in the periodic table could increase. If *alpha* were too big, small atomic nuclei could not exist, because the electrical repulsion of their protons would overwhelm the strong nuclear force binding them together. A value as big as 0.1 would blow carbon apart.

The nuclear reactions in stars are especially sensitive to *alpha*. For fusion to occur, a star's gravity must produce temperatures high enough to force nuclei together despite their tendency to repel one another. If *alpha* exceeded 0.1, fusion would be impossible. A shift of just 4% in *alpha* would alter the energy levels in the nucleus of carbon to such an extent that the production of this element by stars would shut down.

The second experimental problem, which is less easily solved, is that measuring changes in the constants requires high-precision equipment that remains stable long enough to register any changes. Even atomic clocks could detect drifts in the fine-structure constant only over days or, at most, years. If *alpha* changed by more than four parts in 10 to the 15th power over a three-year period, the best clocks would see it. None have. That may sound like an impressive confirmation of constancy, but three years is a cosmic eyeblink. Slow but substantial changes during the long history of the universe would've gone unnoticed.

Fortunately, physicists have found other ways of testing *alpha*. During the 1970s, scientists at the French Atomic Energy Commission noticed something peculiar about the isotopic composition of ore from a uranium mine in Oklo, Gabon: it looked like the waste products of a nuclear reactor. About two billion years ago, they deduced, Oklo must have been the site of a natural reactor.

In 1976, Alexander Shlyakhter of the Petersburg Nuclear Physics Institute in Russia noticed that the ability of a natural nuclear reactor to function depends crucially on the precise energy of a particular state of the samarium nucleus that facilitates the capture of neutrons and that energy depends sensitively on the value of *alpha*. So, if the fine-structure constant had been slightly different, no chain reaction could've occurred. One did occur, though, which implies the constant has not changed by more than one part in 10 to the 8th power over the past two billion years.

In 1962, James Peebles and Robert Dicke of Princeton University first applied similar principles to meteorites. The abundance ratios arising from the radioactive decay of different isotopes in these ancient rocks depend on *alpha*. The most sensitive constraint involves the beta decay of rhenium into osmium. Researchers found that, at the time the rocks formed, *alpha* was within two parts in 10 to the 6th power of its current value. This result is less precise than the Oklo data but it goes back further in time to the origin of our solar system 4.6 billion years ago.

To probe possible changes over even longer time spans, researchers must look to the stars. Light takes billions of years to reach our telescopes from distant astronomical sources. It carries a snapshot of the laws and constants of physics at the time when it started its journey or when it encountered material along the way.

Astronomy first entered the constants story soon after the discovery of quasars in 1965. The idea was simple. Quasars had just been identified as bright sources of light located at huge distances from Earth. Because the path of light from a quasar to us is so long, it inevitably intersects the gaseous outskirts of young galaxies. That gas absorbs the quasar light at particular frequencies, imprinting a bar code of narrow lines onto the quasar spectrum.

Whenever gas absorbs light, electrons within the atoms jump from a low energy state to a higher one. These energy levels are determined by how tightly the atomic nucleus holds the electrons, which depends on the strength of the electromagnetic force between them and therefore on the fine-structure constant *alpha*.

If the constant was different at the time when the light was absorbed or in the particular region of the universe where it happened, then the energy required to lift the electrons would differ from that required today in lab experiments, and the wavelengths of the transitions seen in the spectra would differ. The way the wavelengths change depends critically on the orbital configuration of the electrons. For a given change in *alpha*, some wavelengths shrink, whereas others would increase. The complex pattern of effects is hard to mimic by data-calibration errors, which makes the test astonishingly powerful.

John Barrow and John Webb, who wrote the article this lesson is based on, needed some high-precision lab measurements to compare against the quasar spectra. Initial measurements were done by Anne Thorne and Juliet Pickering of Imperial College London, followed by groups at Lund Observatory in Sweden and the National Institute of Standards and Technology.

Another challenge was that previous observers had used so-called alkali-doublet absorption lines—pairs of absorption lines arising from the same gas, such as carbon or silicon. They compared the spacing between these lines in quasar spectra with lab measurements. However, this method failed to take advantage of one particular phenomenon: a change in *alpha* shifts not just the spacing of atomic energy levels relative to the lowest energy level, or ground state, but also the position of the ground state itself. Consequently, the highest precision observers achieved was only about one part in 10 to the 4th power.

In 1999, John Webb and Victor Flambaum of the University of New South Wales in Sydney came up with a method to take both effects into account and they achieved 10 times higher sensitivity. Moreover, the method allows different species (for instance, magnesium and iron) to be compared, which allows additional cross-checks. Combined with modern telescopes and detectors, the new approach, known as the many-multiplet method, has enabled Webb and Barrow to test the constancy of *alpha* with unprecedented precision.

When embarking on this project, they anticipated establishing that the value of the fine-structure constant long ago was the same as it is today; their contribution would simply be higher precision. To their surprise, the first results showed small but statistically significant differences. Further data confirmed this finding. Based on a total of 128 quasar absorption lines, they found an average increase in *alpha* of close to six parts in a million over the past 6 to 12 billion years.

Extraordinary claims require extraordinary evidence, so Barrow and Webb looked for potential problems with the data or the analysis methods. These kind of uncertainties can be classified into two types: systematic and random. Random uncertainties are easier to understand; they're just that, they're random. They differ for each individual measurement but should average out to be close to zero over a huge sample. Systematic uncertainties, which do not average out, are harder to deal with and they are endemic in astronomy.

Lab experimenters can alter their instrumental setup to minimize systemic errors, but astronomers can't change the universe, and so they're forced to accept that all their methods of gathering data have unavoidable biases. For example, any survey of galaxies will tend to be overrepresented by bright galaxies because they're easier to see. Identifying and neutralizing these biases present a constant challenge.

The first bias Barrow and Webb looked for was a distortion of the wavelength scale the quasar spectral lines were measured against. Such a distortion might conceivably be introduced, for example, during the processing of the quasar data from their raw form at the telescope into a calibrated spectrum. Although a simple linear stretching or compression of the wavelength scale could not precisely mimic a change in *alpha*, even an imprecise mimicry might be enough to explain Barrow and Webb's results. To test for problems of this kind, they substituted calibration data for the quasar data and analyzed them, pretending that they were quasar data. This experiment ruled out simple distortion errors with high confidence.

For more than two years, they put up one potential bias after another, only to rule it out after detailed investigation as being too small to explain the effect. So far, they've identified just one potentially serious source of bias. It concerns the absorption lines produced by the element magnesium.

Each of the three stable isotopes of magnesium absorbs light at a different wavelength, but the three wavelengths are very close to one another, and quasar spectroscopy generally sees the three lines blended as one. Based on lab measurements of the relative abundances of the three isotopes, researchers infer the contribution of each. If these abundances in the young universe differed substantially—as might've happened if the stars that spilled magnesium into their galaxies were, on average, heavier than their counterparts today—these differences could simulate a change in *alpha*.

By 2010, Barrow and Webb completed the analysis of a large amount of new data from what is called the Very Large Telescope, or VLT, operated by the European Southern Observatory and they obtained 153 new measurements. All the data their group had previously analyzed had come from the Keck telescopes on Mauna Kea in Hawaii. For these new VLT data, everything was different—the telescopes, the spectrograph, the detectors and the software used for the initial stages of the data analysis. These VLT data therefore provided a beautiful cross-check with their results from the Keck telescopes.

They thought it was possible that the new data would show no change in *alpha* at all or that they would show the same effect the Keck data did, with *alpha* appearing smaller at higher redshifts. What they actually found was truly astonishing. The new VLT data showed not a smaller value of *alpha* at high redshift but a larger value, larger by just about the same amount as the Keck data are smaller. How can this be? Barrow and Webb's initially suspected they were seeing evidence for systematic problems in both data sets. Add the Keck and VLT samples together, and to a good approximation, the combined sample shows no change in *alpha* with redshift. So, the problem was solved. The constants are really constant after all.

But if that's the explanation, it requires two different systematic effects, one for each telescope, such that both effects are, independently, of the same magnitude but the opposite sign. This isn't impossible, but it's really pretty unlikely. They did discover another curiosity, though. The Keck data cover a largish portion of the sky in the Northern Hemisphere, large enough to ask whether there's any "preferred direction" for the change in *alpha* seen with that sample. So, put another way: Could it be that *alpha* changes not with redshift but with position on the sky?

A simple analysis suggested that might be the case. The VLT is in Chile and, on average, it points to a very different part of the universe than the Keck telescopes do. Another coincidence? Possibly, but that now makes two coincidences. So, what happens when the old Keck and the new VLT samples are merged? The result is positively intriguing: The directional dependence becomes highly significant. Deriving such a result by chance appears to be extremely unlikely. If the result is a fluke, a subset of the data could be generating a rogue result.

Despite extensive attempts, however, they've yet to find a combination of systematic effects in the data that could mimic a spatial dependence. *Alpha* appears to change spatially across, perhaps, the entire observable universe. Any change with time is smaller and is currently below their detection sensitivity. In a study in *Science* in 2020, Webb and Barrow and their colleagues used four different measurements of the fine-structure constant going back 13 billion years and concluded that the spatial variation in *alpha* is real.

The consequences of a variable constant are enormous, though only partially explored. Until quite recently, all attempts to evaluate what happens to the universe if the fine-structure constant changes were unsatisfactory. They amounted to nothing more than assuming that *alpha* became a variable in the same formulas that had been derived assuming it was a constant. According to Barrow and Webb, this is a dubious practice. If *alpha* varies, then its effects must conserve energy and momentum, and they must influence the gravitational field in the universe.

One theory treats *alpha* as not a single number but a so-called scalar field, a dynamic ingredient of nature. This theory makes appealingly simple predictions. Variations in *alpha* of a few parts per million should have a completely negligible effect on the expansion of the universe. But while changes in the fine-structure constant don't affect the expansion of the universe significantly, the expansion affects *alpha*.

According to the theory, changes to *alpha* are driven by imbalances between the electric field energy and magnetic field energy. During the first tens of thousands of years of cosmic history, radiation dominated over charged particles and kept the electric and magnetic fields in balance. As the universe expanded, radiation thinned out, and matter became the dominant constituent of the cosmos. The electric and magnetic energies became unequal, and *alpha* started to increase very slowly, growing as the logarithm of time. About six billion years ago, dark energy took over and accelerated the expansion, making it difficult for all physical influences to propagate through space. So *alpha* became nearly constant again. But even if time variation does take place, it must be small compared with the spatial variation we may now be seeing.

Any theory worthy of consideration doesn't just reproduce observations; it must make novel predictions. Well, Barrow and Webb's theory suggests that varying the fine-structure constant makes objects fall differently. Galileo predicted that bodies in a vacuum fall at the same rate no matter what they're made of—an idea known as the weak equivalence principle. This was demonstrated when Apollo 15 astronaut David Scott dropped a feather and a hammer and saw them hit the lunar dirt at the same time.

But if *alpha* varies, that principle no longer holds exactly. The variations generate a force on all charged particles. The more protons an atom has in its nucleus, the more strongly it will feel this force. If Webb and Barrow's quasar observations are correct, then the accelerations of different materials differ by about one part in 10 to the 14th power, which is much too small to measure currently.

So, where does this flurry of activity leave science as far as *alpha* is concerned? The main focus is on *alpha*, over the other constants of nature, simply because it's possible to build up a statistical sample of measurements, mapping the laws of physics throughout the distant cosmos in greater detail. If *alpha* is susceptible to change, however, other constants should vary as well, making the inner workings of nature more fickle than scientists ever suspected.

11
ATOMS OF SPACE AND TIME

This lesson's topic is loop quantum gravity—a theory that predicts that space and time are made of discrete pieces. Lee Smolin and his colleagues developed the theory of loop quantum gravity while struggling with a long-standing problem in physics: Is it possible to develop a quantum theory of gravity?

A CHALLENGING MERGE

- Formed in the first quarter of the 20th century, quantum theory successfully predicts the properties of atoms and the particles and forces that compose them. The equations of quantum mechanics require that certain quantities, such as the energy of an atom, can come only in specific, discrete units, or quanta.

- In the same decades that quantum mechanics was being formulated, Albert Einstein constructed his general theory of relativity, which is a theory of gravity. In his theory, the gravitational force arises as a consequence of space and time (which together form spacetime) being curved by the presence of matter.

- Quantum theory and Einstein's general theory of relativity have each separately been confirmed by experiment. One problem for merging them is that quantum effects are most prominent at small size scales, whereas general relativistic effects require large masses, so no experiment has been able to combine their domains.

- Along with this hole in the experimental data is a huge conceptual problem: Einstein's general theory of relativity is thoroughly classical, or nonquantum. For physics as a whole to be logically consistent, there has to be a theory that somehow unites quantum mechanics and general relativity.

- This long-sought-after theory is called quantum gravity. Since general relativity deals in the geometry of spacetime, a quantum theory of gravity will also be a quantum theory of spacetime.

- Physicists have developed mathematical procedures for turning a classical theory into a quantum one. Many theoretical physicists and mathematicians have worked on applying those standard techniques to general relativity. Early results were discouraging. Calculations carried out in the 1960s and 1970s seemed to show that quantum theory and general relativity couldn't be combined.

A NEW APPROACH

- In the mid-1980s, Lee Smolin and other theorists decided to reexamine the question of whether quantum mechanics could be combined consistently with general relativity using the standard techniques.

- They knew the negative results from the 1970s had an important loophole. Those calculations assumed that the geometry of space is continuous and smooth. But if this assumption was wrong, the old calculations wouldn't be reliable.

- Smolin and his colleagues began searching for a way to do calculations without assuming that space is smooth and continuous. They kept two key principles of general relativity at the heart of their calculations.

- The first is known as background independence. This principle says that the geometry of spacetime is not fixed. Instead, the geometry is an evolving, dynamic quantity. To find the geometry, one has to solve certain equations that include all the effects of matter and energy.

- The second principle, diffeomorphism invariance, is closely related to background independence. This principle implies that, unlike theories prior to general relativity, one is free to choose any set of coordinates to map spacetime and express the equations. A point in spacetime is defined only by what physically happens at it, not by its location according to some special set of coordinates.

▲ By combining these two principles with the techniques of quantum mechanics, Smolin and his colleagues developed a mathematical language that allowed them to do a computation to determine whether space is continuous or discrete. That calculation revealed, to their delight, that space is quantized.

▲ They had laid the foundations of their theory of loop quantum gravity. The term *loop* arises from how some computations in the theory involve small loops marked out in spacetime.

LOOP QUANTUM GRAVITY **IN ACTION**

▲ Over the years since, the study of loop quantum gravity has grown into a healthy field of research, with many contributors. Their combined efforts give us confidence in the picture of spacetime their theory describes.

▲ This is a quantum theory of the structure of spacetime at the smallest size scales. To explain how the theory works, it is necessary to consider what it predicts for a small region or volume. In dealing with quantum physics, it's essential to specify precisely what physical quantities are to be measured.

Lesson 11 | Atoms of Space and Time

- To do so, consider a region somewhere that's marked out by a boundary, B. The boundary may be defined by some matter, such as a cast-iron shell, or it may be defined by the geometry of spacetime itself, as in the event horizon of a black hole, the boundary where even light can't escape the black hole's gravitational clutches.

- What happens if we measure the volume of the region? What are the possible outcomes allowed by both quantum theory and diffeomorphism invariance? If the geometry of space is continuous, the region could be of any size and the measurement result could be any positive real number; in particular, it could be as close as one wants to zero volume.

- But if the geometry is granular, then the measurement result can come from just a discrete set of numbers, and it can't be smaller than a certain minimum possible volume. The question is like asking how much energy electrons orbiting an atomic nucleus have.

- Classical mechanics predicts that an electron can possess any amount of energy, but quantum mechanics allows only specific energies. The difference is similar to that between the measure of something that flows continuously, like the 19th-century conception of water, and something that can be counted, like the atoms in that water.

- The theory of loop quantum gravity predicts that space is like atoms: There is a discrete set of numbers that the volume-measuring experiment can return. Another quantity we can measure is the area of the boundary B. Again, calculations using the theory return an unambiguous result: The area of the surface is discrete as well. In other words, space is not continuous. It comes only in specific quantum units of area and volume.

THE PLANCK LENGTH

- The possible values of volume and area are measured in units of a quantity called the Planck length. This length is related to the strength of gravity, the size of quanta and the speed of light. It measures the scale at which the geometry of space is no longer continuous.

- The Planck length is very small: 10^{-33} centimeter. The smallest possible nonzero area is about a square Planck length, or 10^{-66} centimeter squared. The smallest nonzero volume is approximately a cube with edges of Planck length, 10^{-99} centimeter cubed.

- The theory predicts that there are about 10^{99} atoms of volume in every cubic centimeter of space. The quantum of volume is so tiny that there are more such quanta in a cubic centimeter than there are cubic centimeters in the visible universe.

DIAGRAMS

- This lesson turns next to examine what else the theory can tell us about spacetime. To start with, what do these quantum states of volume and area look like? Is space made up of many tiny cubes or spheres?

- The answer is no—it's not that simple. However, it is possible to draw diagrams that represent the quantum states of volume and area. To those working in this field, these diagrams are beautiful because of their connection to an elegant branch of mathematics.

Lesson 11 | Atoms of Space and Time

▲ To see how these diagrams work, imagine a lump of space shaped like a cube. In the diagrams, this would be depicted as a dot, which represents the volume, with six lines sticking out, each of which represents one of the cube's faces. A number next to the dot to specifies the quantity of volume, and on each line, a number to specify the area of the face that the line represents.

▲ Next, imagine a pyramid on top of the cube. These two polyhedral shapes, which share a common face, would be depicted as two dots (two volumes) connected by one of the lines (the face that joins the two volumes). The cube has five other faces (five lines sticking out), and the pyramid has four (four lines sticking out).

▲ More complicated arrangements involving polyhedra other than cubes and pyramids can be depicted with these dot-and-line diagrams: Each polyhedron of volume becomes a dot, or node, and each flat face of a polyhedron becomes a line. The lines join the nodes in the way that the faces join the polyhedra together. Mathematicians call these line diagrams graphs.

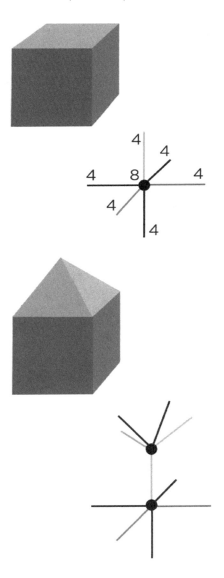

- In loop quantum gravity, we throw away the drawings of polyhedra and just keep the graphs. The mathematics that describes the quantum states of volume and area gives us a set of rules for how the nodes and lines can be connected and what numbers can go where in a diagram.

- Every quantum state corresponds to one of these graphs, and every graph that obeys the rules corresponds to a quantum state. The graphs are a convenient shorthand for all the possible quantum states of space.

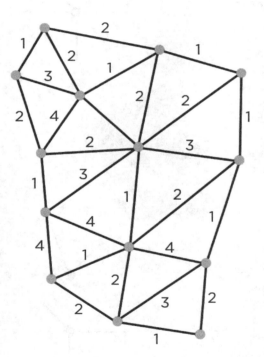

- The graphs are a better representation of the quantum states than the polyhedra are. In particular, some graphs connect in strange ways that can't be converted into a tidy picture of polyhedra. For example, whenever space is curved, the polyhedra will not fit together properly in any drawing we could do, yet we can still draw a graph.

- Indeed, we can take a graph and from it calculate how much space is distorted. Because the distortion of space is what produces gravity, this is how the diagrams form a quantum theory of gravity.

Lesson 11 | Atoms of Space and Time

◢ For simplicity's sake, we often draw the graphs in two dimensions. But it's actually better to imagine them filling three-dimensional space since that's what they represent. Yet there's a conceptual trap here: The lines and nodes of a graph don't live at specific locations in space.

◢ Each graph is defined only by the way its pieces connect and how they relate to well-defined boundaries such as boundary B. The continuous, three-dimensional space one might imagine the graphs occupying does not exist as a separate entity. All that exist are the lines and nodes; they are space, and the way they connect defines the geometry of space.

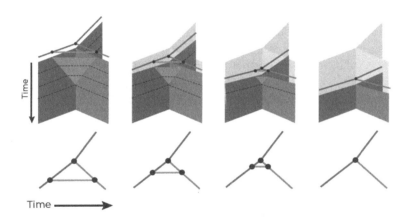

SPIN NETWORKS

◢ These graphs are called spin networks because the numbers on them are related to quantities called spins. Roger Penrose of the University of Oxford first proposed in the early 1970s that spin networks might play a role in theories of quantum gravity. In 1994, Smolin and his colleagues found that precise calculations confirmed Penrose's intuition.

- The individual nodes and edges of the diagrams represent extremely small regions of space: A node is typically a volume of about one cubic Planck length, and a line is typically an area of about one square Planck length. But in principle, nothing limits how big and complicated a spin network can be.

- If we could draw a detailed picture of the quantum state of our universe—the geometry of its space, as curved and warped by the gravitation of galaxies and black holes and everything else—it would be a gargantuan spin network of unimaginable complexity, with approximately 10^{184} nodes.

- These spin networks describe the geometry of space. But how do we represent particles and fields occupying positions and regions of space? Particles such as electrons correspond to certain types of nodes, which are represented by adding more labels on nodes. Fields, such as the electromagnetic field, are represented by additional labels on the lines of the graph.

- We represent particles and fields moving through space by these labels moving in discrete steps on the graphs. Particles and fields aren't the only things that move around, though. According to general relativity, the geometry of space changes in time. The bends and curves of space change as matter and energy move, and waves can pass through it like ripples on a lake.

- In loop quantum gravity, these processes are represented by changes in the graphs. They evolve in time by a succession of certain "moves" in which the connectivity of the graphs changes.

PHENOMENA AND PROBABILITIES

- When physicists describe phenomena quantum-mechanically, they compute probabilities. We do the same when we apply loop quantum gravity theory to describe phenomena, whether it's particles and fields moving on the spin networks or the geometry of space itself evolving in time.

Lesson 11 | Atoms of Space and Time

- Work by many people over the past few decades has revealed an elegant set of rules for computing the probabilities of the different moves by which spin networks change in time. These rules express the quantum version of Einstein's equations of general relativity. With them, we have a well-defined procedure for computing the probability of any process that can occur in a world that obeys the rules of our theory.

- To discover the precise rules for computing probabilities, physicists had to follow Einstein in shifting the perspective from space to spacetime. The spin networks that represent space in loop quantum gravity theory accommodate the concept of spacetime by becoming what we call spin foams.

- With the addition of another dimension—time—the lines of the spin networks grow to become two-dimensional surfaces and the nodes grow to become lines. Transitions where the spin networks change are now represented by nodes where the lines meet in the foam.

- In the spacetime way of looking at things, a snapshot at a specific time is like a slice cutting across the spacetime. Taking such a slice through a spin foam produces a spin network.

- But it would be wrong to think of such a slice as moving continuously, like a smooth flow of time. Instead, just as space is defined by a spin network's discrete geometry, time is defined by the sequence of distinct moves that rearrange the network.

- In this way, time also becomes discrete. Time flows not like a river but like the ticking of a clock, with "ticks" that are about as long as the Planck time: 10^{-43} second. More precisely, time in our universe flows by the ticking of innumerable clocks. In a sense, at every location in the spin foam where a quantum "move" takes place, a clock at that location has ticked once.

CONCLUSION

- Loop quantum gravity has opened up a new window on deep cosmological questions such as those relating to the origins of our universe. We can use the theory to study the earliest moments of time just after the big bang.

- General relativity predicts that there was a first moment of time. Models of the very early universe based on loop quantum gravity, however, indicate that the big bang is actually a big bounce. Before the bounce, the universe was rapidly contracting.

- Theorists are now hard at work developing predictions for the early universe that may be testable in future cosmological observations. Perhaps this will lead to evidence of the time before the big bang.

ABOUT THIS LESSON

This lesson was adapted from the article "Atoms of Space and Time" by Lee Smolin of the Perimeter Institute for Theoretical Physics. Smolin's books, including *Time Reborn*, probe philosophical issues raised by research in physics and cosmology.

Lesson 11 Transcript

ATOMS OF SPACE AND TIME

This lesson was adapted from an article by Lee Smolin of the Perimeter Institute for Theoretical Physics. Smolin's books, including *Time Reborn*, probe philosophical issues raised by research in physics and cosmology.

Little more than 100 years ago most people—including most scientists—thought of matter as continuous. Since ancient times some philosophers had speculated that if matter were broken up into small enough bits, it might turn out to be made up of very tiny atoms, but few thought the existence of atoms could ever be proved. Today we've imaged individual atoms and studied the particles composing them. The granularity of matter is real.

But in recent decades, physicists have asked whether space is also made of discrete pieces. Is it continuous or is it more like a piece of cloth, woven out of individual fibers? If we could probe to size scales that were small enough, would we see "atoms" of space, pieces of volume that can't be broken into anything smaller? They're asking the same question about time: Does nature change continuously, or does the world evolve in a series of very tiny steps?

A theory with the strange name of "loop quantum gravity" predicts that space and time are indeed made of discrete pieces. The picture revealed by the framework of this theory is both simple and beautiful. The theory has deepened our understanding of puzzling phenomena having to do with black holes and the big bang. Best of all, it's possible that current experiments might be able to detect signals of the atomic structure of spacetime—if such a structure actually exists—potentially in the near future.

Lee Smolin and his colleagues developed the theory of loop quantum gravity while struggling with a long-standing problem in physics: Is it possible to develop a quantum theory of gravity? To explain why this is such an important question—and what it has to do with the granularity of space and time—let's first talk a bit about quantum theory and the theory of gravity.

The theory of quantum mechanics was formulated in the first quarter of the 20th century and was closely connected with the confirmation that matter is indeed made of atoms. The equations of quantum mechanics require that certain quantities, such as the energy of an atom, can come only in specific, discrete units, or quanta. Quantum theory successfully predicts the properties of atoms and the particles and forces that compose them. It underlies our understanding of chemistry, atomic and subatomic physics, electronics and even some biology.

Now, in the same decades that quantum mechanics was being formulated, Albert Einstein constructed his general theory of relativity, which is a theory of gravity. In his theory, the gravitational force arises as a consequence of space and time (which together form spacetime) are curved by the presence of matter.

So, here's a loose analogy: Imagine a bowling ball is placed on a rubber sheet along with a marble that's rolling around nearby. The balls could represent the Sun and the Earth and the sheet is space. The bowling ball creates a deep indentation in the rubber sheet, and the slope of this indentation causes the marble to roll toward the larger ball, as if some force were pulling it in that direction. Similarly, any piece of matter or concentration of energy distorts the geometry of spacetime, causing other particles and light rays to be deflected toward it, and this the phenomenon we call gravity.

Quantum theory and Einstein's general theory of relativity have each separately been fantastically well confirmed by all kinds of experiments. One problem for merging them is that quantum effects are most prominent at very small size scales, whereas general relativistic effects require large masses, so no experiment has been able to combine their domains.

Along with this hole in the experimental data is a huge conceptual problem: Einstein's general theory of relativity is thoroughly classical, or nonquantum.

Lesson 11 Transcript | Atoms of Space and Time

For physics as a whole to be logically consistent, there has to be a theory that somehow unites quantum mechanics and general relativity. This long-sought-after theory is called quantum gravity. Since general relativity deals in the geometry of spacetime, a quantum theory of gravity will also be a quantum theory of spacetime.

Physicists have developed a whole collection of mathematical procedures for turning a classical theory into a quantum one. Many theoretical physicists and mathematicians have worked on applying those standard techniques to general relativity. Early results were pretty discouraging. Calculations carried out in the 1960s and 1970s seemed to show that quantum theory and general relativity simply couldn't be combined. So, something fundamentally new seemed to be required, such as maybe additional postulates or principles not included in either theory, or maybe new particles or fields or other entities of some kind.

To avoid spoiling the successful predictions in both quantum theory and general relativity, the exotica contained in the full theory would remain hidden from experiment except in the extraordinary circumstances where both quantum theory and general relativity are expected to have large effects. Many different approaches along these lines have been tried, with names like twistor theory, supergravity and string theory. But after many years of study, none of these ideas has led to unambiguous predictions that can be tested experimentally. So, many physicists have started to doubt whether quantum theory and general relativity are compatible after all.

In the mid-1980s, Lee Smolin and other theorists decided to reexamine the question of whether quantum mechanics could be combined consistently with general relativity using the standard techniques. They knew the negative results from the 1970s had an important loophole. Those calculations assumed that the geometry of space is continuous and smooth, no matter how minutely we examine it, just as people had understood matter before the discovery of atoms. But if this assumption was wrong, the old calculations wouldn't be reliable.

So, Smolin and his colleagues began searching for a way to do calculations without assuming that space is smooth and continuous. They insisted on not making any assumptions beyond the experimentally well tested principles of general relativity and quantum theory. In particular, they kept two key principles of general relativity at the heart of their calculations. The first is what's known as background independence. This principle says that the geometry of spacetime is not fixed. Instead, the geometry is an evolving, dynamic quantity. To find the geometry, one has to solve certain equations that include all the effects of matter and energy.

The second principle, known by the imposing name diffeomorphism invariance, is closely related to background independence. This principle implies that, unlike theories prior to general relativity, one is free to choose any set of coordinates to map spacetime and express the equations. A point in spacetime is defined only by what physically happens at it, not by its location according to some special set of universal coordinates.

By combining these two principles with the techniques of quantum mechanics, Smolin and his colleagues developed a mathematical language that allowed them to do a computation to determine whether space is continuous or discrete. That calculation revealed, to their delight, that space is quantized. They'd laid the foundations of their theory of loop quantum gravity. The term *loop*, by the way, arises from how some computations in the theory involve small loops marked out in spacetime.

The calculations have been redone by a number of physicists and mathematicians using a range of methods. Over the years since, the study of loop quantum gravity has grown into a healthy field of research, with many contributors around the world; their combined efforts give us confidence in the picture of spacetime their theory describes. This is a quantum theory of the structure of spacetime at the smallest size scales. So to explain how the theory works, we need to consider what it predicts for a small region or volume. In dealing with quantum physics, it's essential to specify precisely what physical quantities are to be measured.

To do so, let's consider a region somewhere that's marked out by a boundary, identified as B. The boundary may be defined by some matter, such as a cast-iron shell, or it may be defined by the geometry of spacetime itself, as in the event horizon of a black hole; that is the boundary where even light can't escape the black hole's gravitational clutches.

So, what happens if we measure the volume of the region? What are the possible outcomes allowed by both quantum theory and diffeomorphism invariance? If the geometry of space is continuous, the region could be of any size and the measurement result could be any positive real number; in particular, it could be as close as one wants to zero volume. But if the geometry is granular, then the measurement result can come from just a discrete set of numbers and it can't be smaller than a certain minimum possible volume.

The question is like asking how much energy electrons orbiting an atomic nucleus have. Classical mechanics predicts that an electron can possess any amount of energy, but quantum mechanics allows only specific energies. The difference is similar to that between the measure of something that flows continuously, like the 19th-century conception of water, and something that can be counted, like the atoms in that water.

The theory of loop quantum gravity predicts that space is like atoms: There's a discrete set of numbers that the volume-measuring experiment can return. Another quantity we can measure is the area of the boundary B. Again, calculations using the theory return an unambiguous result: The area of the surface is discrete as well. In other words, space is not continuous. It comes only in specific quantum units of area and volume.

The possible values of volume and area are measured in units of a quantity called the Planck length. This length is related to the strength of gravity, the size of quanta and the speed of light. It measures the scale at which the geometry of space is no longer continuous. The Planck length is very small: 10 to the negative 33rd centimeter. The smallest possible non-zero area is about a square Planck length, or 10 negative-66th centimeter squared. The smallest non-zero volume is approximately a cube with edges of Planck length, 10 to the negative-99th centimeter cubed.

The theory predicts that there are about 10 to the 99th power atoms in every cubic centimeter of space. The quantum of volume is so tiny that there are more such quanta in a cubic centimeter than there are cubic centimeters in the visible universe.

So, what else does this theory tell us about spacetime? To start with, what do these quantum states of volume and area look like? Is space made up of a lot of little cubes or a lot of little spheres? The answer is no; it's not that simple. Nevertheless, we can draw diagrams that represent the quantum states of volume and area. To those working in this field, these diagrams are beautiful because of their connection to an elegant branch of mathematics.

To see how these diagrams work, imagine we've got a lump of space shaped like a cube. In our diagrams, we'd depict this cube as a dot, which represents the volume, with six lines sticking out, each of which represents one of the cube's faces. We write a number next to the dot to specify the quantity of volume, and on each line, we write a number to specify the area of the face that the line represents. Next, suppose we put a pyramid on top of the cube. These two polyhedral shapes, which share one common face, would be depicted as two dots (representing two volumes) connected by one of the lines (which is the face that joins the two volumes), and the cube has five other faces (or five lines sticking out), and the pyramid has four (or four lines sticking out).

It's clear how more complicated arrangements involving polyhedra other than cubes and pyramids could be depicted with these dot-and-line diagrams. Each polyhedron of volume becomes a dot, or node and each flat face of a polyhedron becomes a line. The lines join the nodes in such a way that the faces join the polyhedra together. Mathematicians call these line diagrams graphs.

Now, in loop quantum gravity, we throw away the drawings of polyhedra and just keep the graphs. The mathematics that describes the quantum states of volume and area gives us a set of rules for how the nodes and lines can be connected and what numbers can go where in a diagram. Every quantum state corresponds to one of these graphs and every graph that obeys the rules corresponds to a quantum state. The graphs are a convenient shorthand for all the possible quantum states of space.

The graphs are a better representation of the quantum states than the polyhedra are. In particular, some graphs connect in strange ways that can't be converted into a tidy picture of polyhedral shapes. For example, wherever space is curved, the polyhedra will not fit together properly in any drawing we could do, yet we can still draw a graph. Indeed, we can take a graph and from it calculate how much space is distorted. Because the distortion of space is what produces gravity, this is how the diagrams form a quantum theory of gravity.

For simplicity's sake, we often draw the graphs in two dimensions. But it's actually better to imagine them filling three-dimensional space because that's what they represent. Yet there's a conceptual trap here; the lines and nodes of a graph don't live at specific locations in space. Each graph is defined only by the way its pieces connect and how they relate to well-defined boundaries such as the boundary B. The continuous, three-dimensional space you're imagining the graphs occupy does not exist as a separate entity. All that exist are the lines and nodes; they are space, and the way they connect defines the geometry of space.

These graphs are called spin networks because the numbers on them are related to quantities called spins. Roger Penrose of the University of Oxford first proposed in the early 1970s that spin networks might play a role in theories of quantum gravity. In 1994, Smolin and his colleagues found that precise calculations confirmed Penrose's intuition.

The individual nodes and edges of the diagrams represent extremely small regions of space. A node is typically a volume of about one cubic Planck length and a line is typically an area of about one square Planck length. But in principle, nothing limits how big and complicated a spin network can be. If we could draw a detailed picture of the quantum state of our universe—the geometry of its space, as curved and warped by the gravitation of galaxies and black holes and everything in it—it would be a gargantuan spin network of unimaginable complexity, with approximately 10 to the 184th power nodes.

These spin networks describe the geometry of space. But how do we represent particles and fields occupying positions and regions of space? Particles such as electrons correspond to certain types of nodes, which are represented by adding more labels on the nodes. Fields, such as the electromagnetic field, are represented by additional labels on the lines of the graph. We represent particles and fields moving through space by these labels moving in discrete steps on the graphs.

Particles and fields aren't the only things that move around, though. According to general relativity, the geometry of space changes in time. The bends and curves of space change as matter and energy move and waves can pass through it like ripples on a lake.

In loop quantum gravity, these processes are represented by changes in the graphs. They evolve in time by a succession of certain moves in which the connectivity of the graphs change. When physicists describe phenomena quantum-mechanically, they compute probabilities. We do the same when we apply loop quantum gravity theory to describe phenomena, whether it's particles and fields moving on the spin networks or the geometry of space itself evolving in time.

Work by many people over the past few decades has revealed an elegant set of rules for computing the probabilities of the different moves by which spin networks change in time. These rules express the quantum version of Einstein's equations of general relativity. With them, we have a well-defined procedure for computing the probability of any process that can occur in a world that obeys the rules of our theory.

To discover the precise rules for computing probabilities, physicists had to follow Einstein in shifting the perspective from space to spacetime. The spin networks that represent space in loop quantum gravity theory accommodate the concept of spacetime by becoming what we call spin foams. With the addition of another dimension—time—the lines of the spin network grow to become two-dimensional surfaces and the nodes grow to become lines. Transitions where the spin networks change are now represented by nodes where the lines meet in the foam.

In the spacetime way of looking at things, a snapshot at a specific time is like a slice cutting across spacetime. Taking such a slice through a spin foam produces a spin network. But it would be wrong to think of such a slice as moving continuously, like a smooth flow of time. Instead, just as space is defined by a spin network's discrete geometry, time is defined by the sequence of distinct moves that rearrange the network.

In this way, time also becomes discrete. Time flows not like a river but like the ticking of a clock, with ticks that are about as long as the Planck time: 10 to the negative 43rd second. Or, more precisely, time in our universe flows by the ticking of innumerable clocks—in a sense, at every location in the spin foam where a quantum move takes place, a clock at that location has ticked once.

Okay, so we've outlined what loop quantum gravity has to say about space and time at the Planck scale, but we can't verify the theory directly by examining spacetime on that scale. It's just too small. So how can we test the theory? An important test is whether we can derive classical general relativity as an approximation to loop quantum gravity. In other words, if the spin networks are like the threads woven into a piece of cloth, this is analogous to asking whether we can compute the right elastic properties for a sheet of the material by averaging over thousands of threads. Similarly, when averaged over many Planck lengths, do spin networks describe the geometry of space and its evolution in a way that agrees roughly with the smooth cloth of Einstein's classical theory?

This is a challenging problem, but with the right rules for computing probabilities, physicists have made a lot of progress. The work of several groups has made it possible to assert that when the elements of the discrete geometry are large, compared with the Planck scale, they behave in a way that is a good approximation of Einstein's equations of general relativity. In the same approximation, it's been shown that gravitons—which are the quanta associated with gravitational waves—travel and scatter from one another precisely as Einstein's general theory of relativity dictates.

Another fruitful test is to see what loop quantum gravity has to say about one of the long-standing mysteries of gravitational physics and quantum theory: The thermodynamics of black holes; in particular, their entropy, which is related to disorder. Physicists have computed predictions regarding black hole thermodynamics using a hybrid It is an approximate theory in which matter is treated quantum-mechanically but spacetime is not.

Specifically, in the 1970s, Jacob Bekenstein at the Hebrew University of Jerusalem inferred that black holes must have an entropy proportional to their surface area. Shortly after, Stephen Hawking of the University of Cambridge deduced that black holes, particularly small ones, must emit radiation.

To do the calculation in loop quantum gravity, we pick the boundary B to be the event horizon of a black hole. When we analyze the entropy of the relevant quantum states, we get precisely Bekenstein's prediction. Similarly, the theory reproduces Hawking's prediction of black hole radiation. In fact, it makes further predictions for the fine structure of Hawking radiation.

An experimental test of any quantum theory of gravity seems like an immense technological challenge. The problem is that the characteristic effects described by the theory become significant only at the Planck scale. But a few imaginative researchers have thought up new ways to test the predictions of loop quantum gravity. These methods depend on the propagation of light across the universe. When light moves through a medium, its wavelength suffers some distortions, leading to effects such as bending in water and the separation of different wavelengths, or colors through a prism. These effects may also occur for light and particles moving through the discrete space described by a spin network.

Unfortunately, the magnitude of the effects is proportional to the ratio of the Planck length to the wavelength. For visible light, this ratio is smaller than 10 to the negative 28th power. So, even for the most powerful cosmic rays ever observed, it's about one billionth. For any radiation we can observe, the effects of the granular structure of space are very small. But these effects accumulate when light travels a long distance and we detect light and particles that come from billions of light-years away, from events such as gamma-ray bursts.

Gamma-ray bursts contain photons of very different energies. According to Einstein's special theory of relativity, all photons travel at the universal speed of light. Consequently, the photons from a burst should arrive in the order they're emitted. But a possible consequence of the light traveling through a discrete spacetime is to modify this law so that a photon's speed depends very slightly on its energy.

This would imply that the quantum structure of spacetime comes into conflict with Einstein's special theory of relativity. This conflict is very small on usual scales but becomes detectable when we study light that's traveled across enormous distances, over which time the effect is amplified so that more energetic photons tend to arrive earlier than their less energetic siblings.

So far, observations confirm the validity of the relativity principle even at scales of quantum geometry. Still, loop quantum gravity has opened up a new window on deep cosmological questions such as those relating to the origins of our universe. We can use the theory to study the earliest moments of time just after the big bang.

General relativity predicts that there was a first moment of time. Models of the very early universe based on loop quantum gravity, however, indicate that the big bang is actually a big bounce; before the bounce the universe was rapidly contracting. Theorists are now hard at work developing predictions for the early universe that may be testable in future cosmological observations. It's not impossible that in our lifetimes we could see evidence of the time before the big bang.

Research on loop quantum gravity also addresses the unification of gravity with the other forces found in nature. It's even possible to incorporate extra dimensions and supersymmetry in the theory, if needed. But as in string theory, no principle has emerged that would predict a unique unification of gravity within particle physics.

Many open questions remain in loop quantum gravity. While there's now good evidence that general relativity emerges as an approximation to the theory within certain limits, we'd like to better understand how robust this property is. And we must know what modifications of relativity theory are implied, as these could lead to observable effects.

Loop quantum gravity occupies a very important place in the development of physics. It's arguably the quantum theory of general relativity because it makes no extra assumptions beyond the basic principles of quantum theory and relativity theory. The remarkable departure it makes—proposing a discontinuous spacetime described by spin networks and spin foams—emerges from the mathematics of the theory itself, rather than being inserted as some kind of ad hoc fix.

Still, everything we've discussed is theoretical. It could be that despite all we've seen here, space really is continuous, no matter how small the scale we probe. Since this is science, in the end experiments will decide.

12

COULD TIME END?

In our experience, time never really ends. But will that always be the case? Might there come a point sometime in the future when there's no "after"? Some modern physics suggests the answer is yes. All activity would cease, and there would be no renewal or recovery. The end of time would be the end of endings. For now, though, it is the topic of this lesson.

Lesson 12 | Could Time End?

AN **UNANTICIPATED** PREDICTION

▲ This prospect was an unanticipated prediction of Albert Einstein's general theory of relativity, which provides our modern understanding of gravity. Before that theory, most physicists and philosophers thought time was a universal drumbeat, a steady rhythm that the cosmos marches to, never varying, wavering, or stopping.

▲ However, Einstein showed that the universe is more like a big polyrhythmic jam session. Time can slow down, or stretch out, or let it rip. When we feel the force of gravity, we're feeling time's rhythmic improvisation; falling objects are drawn to places where time passes more slowly. Time not only affects what matter does but also responds to what matter is doing. When things get out of hand, though, time can go up in smoke.

▲ The moments when that happens are known as singularities. A singularity is any boundary of time, be it beginning or end. The best known is the big bang, the instant 13.8 billion years ago when our universe—and, with it, time—burst into existence and began expanding.

▲ Relativity says time expires inside black holes while carrying on in the universe at large. If a living being fell into one, the being would be torn to shreds, and the remains would eventually hit a singularity at the center of the hole. The being's time line would end. This would be complete death without rebirth.

- It took physicists decades to accept that relativity theory predicts something so unsettling. Singularities are arguably the main reason physicists seek a unified theory of physics that would merge Einstein's brainchild of relativity with quantum mechanics to create a quantum theory of gravity. They do so partly in the hope they might explain singularities away.

THE ALTERNATIVE TO TIME ENDING

- Time's end is hard to imagine, but time's not ending may be equally paradoxical. Well before Einstein came along, philosophers through the ages had debated whether time could be mortal.

- Aristotle, for instance, argued that time can't have a beginning or an end. Every moment is both the end of an era and the start of something new; every event is both the outcome of something and the cause of something else.

- The University of Oxford philosopher Richard Swinburne has asserted, "It is not logically possible for time to have an end." But if time cannot end, then the universe must be infinitely long-lived, and all the riddles posed by the notion of infinity come rushing in. Philosophers have thought it absurd that infinity could be anything but a mathematical idealization.

- The triumph of the big bang theory and the discovery of black holes seemed to settle the question. The universe is shot through with singularities and could suffer a distressing variety of temporal cataclysms. But when it comes to figuring out what singularities actually are, the answer's not so clear.

- At the big bang singularity, relativity theory says that the precursors of every single galaxy we see were squashed into a single mathematical point—a true point of zero size.

- Likewise, in a black hole, every single particle of a hapless astronaut is compacted into an infinitesimal point. In both cases, calculating the density means dividing by zero volume, yielding infinity. Other types of singularities don't involve infinite density but an infinite something else.

- Modern physicists take infinity as a sign they've pushed a theory too far. Nearly all physicists presume that cosmic singularities actually have a finite, if high, density. Relativity theory errs because it fails to capture some important aspect of gravity or matter that comes into play near singularities and keeps the density under control.

NEW APPROACHES

- To figure out what goes on will take a more encompassing theory—a quantum theory of gravity. Physicists are still working on such a theory, but they figure that it will incorporate the central insight of quantum mechanics: that matter, like light, has wavelike properties.

- These properties should smear the putative singularity into a small wad, rather than a point, and thereby banish the divide-by-zero error. If so, time may not, in fact, end.

- However, physicists argue it both ways. Some think time does end. The trouble with this option is that the known laws of physics operate within time and describe how things move and evolve. Time's end points would have to be governed not just by a new law of physics but by a new type of law of physics, one that avoids temporal concepts such as motion and change in favor of timeless ones such as geometric elegance.

- A portion of quantum gravity researchers think that time stretches on forever, with no beginning or end. In their view, the big bang was simply a dramatic transition in the eternal life of the universe.

OTHER WAYS OF THINKING

- Some people conclude that science can never resolve whether time ends. For them, the boundaries of time are also the boundaries of reason and empirical observation. Others, however, think figuring out the boundaries of time simply requires some fresh thinking. For example, physicist Gary Horowitz says, "Quantum gravity should be able to provide a definite answer."

- Just as life emerges out of lifeless molecules that organize themselves, time might emerge from some timeless stuff that brings itself to order. A temporal world is a highly structured one. Time tells us when events occur, for how long, and in what order. Perhaps this structure was not imposed from the outside but arose from within. What can be made can be unmade. When the structure crumbles, time ends.

- By this thinking, time's demise is no more paradoxical than the disintegration of any other complex system. One by one, time loses its features and passes through the twilight from existence to nonexistence.

- The first to go might be its unidirectionality— its "arrow" pointing from past to future. Physicists have recognized since the mid-19th century that the arrow is a property not of time per se but of matter. Time is inherently bidirectional; the arrow we perceive is simply the natural degeneration of matter from order to chaos.

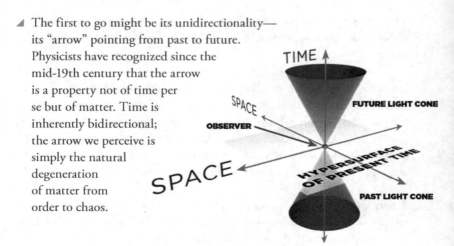

- If this trend keeps up, the universe will approach a state of equilibrium in which it can't possibly get any messier. Individual particles will continue to reshuffle themselves, but the universe as a whole will cease to change. Any surviving clocks will jiggle in both directions, and the future will become indistinguishable from the past.

- Another feature time might lose could be the concept of duration. Time as we know it comes in amounts such as seconds, days, and years. If it didn't, we'd be able to tell that events occurred in chronological order but not how long they lasted. That scenario is what University of Oxford physicist Roger Penrose presents in his book *Cycles of Time: An Extraordinary New View of the Universe*.

TIME AS A DIMENSION OF SPACE

- Even if duration becomes meaningless, time isn't dead quite yet. It still dictates that events unfold in a sequence of cause and effect that is the same for all observers. In this respect, time is different from space.

- Two events that are adjacent within time—like a person typing on a keyboard and letters appearing on their computer screen—are inextricably linked. But two objects that are adjacent within space—the keyboard and a nearby piece of paper—might have nothing to do with each other. Spatial relations simply don't have the same inevitability that temporal ones do.

- Under certain conditions, though, time could lose even this basic ordering function and become just another dimension of space. This idea goes back to the 1980s, when Stephen Hawking and James Hartle sought to explain the big bang as the moment when time and space became differentiated.

THE **HOLOGRAPHIC** PRINCIPLE

- Even if we can't define duration or causal relations, we can still label events by the time they occurred and lay them out on a time line. Several groups have made progress on how time might be stripped of this last remaining feature by studying what happens to it at black hole singularities using one of the most powerful ideas of string theory, known as the holographic principle.

- A hologram is a special type of image that evokes a sense of depth. Though flat, the hologram is patterned to make it look as though a solid object is floating in front of you in three-dimensional space.

- The holographic principle holds that our entire universe is like a holographic projection. A complex system of interacting quantum particles can evoke a sense of depth—that is to say, a spatial dimension that doesn't exist in the original system.

- However, the converse is not true. Not every image is a hologram; it must be patterned in just the right way. If a hologram is scratched, its illusion is spoiled. Likewise, not every particle system gives rise to a universe like ours. The system must be patterned just so.

- If the system initially lacks the necessary regularities and then develops them, the spatial dimension pops into existence. If the system reverts to disorder, the dimension disappears where it came from.

- With that in mind, imagine the collapse of a star to a black hole. The star looks three-dimensional to us but corresponds to a pattern in some two-dimensional particle system. As its gravity intensifies, the corresponding planar system jiggles with increasing fervor. When a singularity forms, order breaks down completely.

- The process is analogous to the melting of an ice cube: The water molecules go from a regular crystalline arrangement to the disordered jumble of a liquid. The third dimension literally melts away. As it goes, so does time.

- If an astronaut falls into a black hole, the time on the astronaut's watch depends on their distance from the center of the hole, which is defined within the melting spatial dimension. As that dimension disintegrates, the watch starts to spin uncontrollably, and it becomes impossible to say that events occur at specific times or objects reside in specific places.

- In practice, this means that space and time no longer give structure to the world. Attempted measurements of the objects' positions will show that they appear to reside in more than one place.

- Spatial separation means nothing to them; they jump from one place to another without crossing the intervening distance. In fact, that's how the imprint of a hapless astronaut who passes the black hole's point of no return, its event horizon, can get back out. "If space and time do not exist near a singularity, the event horizon is no longer well defined," Gary Horowitz says.

- In other words, string theory doesn't just smear out the putative singularity, replacing the errant point with something more palatable while leaving the rest of the universe much the same. Instead, it reveals a broader breakdown of the concepts of space and time, the effects of which persist far from the singularity itself.

◢ The theory still requires a primal notion of time in the particle system. Scientists are still trying to develop a notion of dynamics that doesn't presuppose time at all. Until then, time clings stubbornly to life. It's so deeply ingrained in physics that scientists have yet to imagine its final and total disappearance.

ABOUT THIS LESSON

This lesson was adapted from the article "Could Time End?" by George Musser, a contributing editor for *Scientific American* and author of *Spooky Action at a Distance*. The original article won the 2011 Science Communication Award from the American Institute of Physics.

Lesson 12 Transcript

COULD TIME END?

This lesson was adapted from an article by George Musser, a contributing editor for *Scientific American* and author of *Spooky Action at a Distance*.

In our experience, time never really ends. But will that always be the case? Might there come a point sometime in the future when there's no *after*? Some modern physics suggests the answer is yes. All activity would cease, and there would be no renewal or recovery, no next things. The end of time would be the end of endings.

This prospect was an unanticipated prediction of Einstein's general theory of relativity, which provides our modern understanding of gravity. Before that theory, most physicists and philosophers thought time was a universal drumbeat, a steady rhythm that the cosmos marches to, never varying, wavering or stopping. Albert Einstein showed that the universe is more like a big polyrhythmic jam session. Time can slow down, or stretch out, or let it rip.

When we feel the force of gravity, we're feeling time's rhythmic improvisation; falling objects are drawn to places where time passes more slowly. Time not only affects what matter does but also responds to what matter is doing, like drummers and dancers firing one another up into a rhythmic frenzy. When things get out of hand, though, time can go up in smoke like an overexcited drummer who spontaneously combusts.

The moments when that happens are known as singularities. The term actually refers to any boundary of time, be it beginning or end. The best known is the big bang, the instant 13.8 billion years ago when our universe—and, with it, time—burst into existence and began expanding.

Relativity says time expires inside black holes while carrying on in the universe at large. Black holes have a well-deserved reputation for destructiveness, but they're even worse than you might think. If you fell into one, you'd not only get torn to shreds, but your remains would eventually hit a singularity at the center of the black hole and your time line would end. No new life would emerge from your ashes; your molecules wouldn't get recycled. Like a character reaching the last page of a novel, you'd suffer not mere death but existential apocalypse.

It took physicists decades to accept that relativity theory predicts something so unsettling as death without rebirth. To this day, they aren't quite sure what to make of it. Singularities are arguably the main reason physicists seek a unified theory of physics that would merge Einstein's brainchild of relativity with quantum mechanics to create a quantum theory of gravity. They do so partly in the hope that they might explain singularities away.

Still, we need to be careful what we wish for. Time's end is hard to imagine, but time's not ending may be equally paradoxical. Well before Einstein came along, philosophers through the ages had debated whether time could be mortal. Immanuel Kant considered the issue to be an "antinomy"—something you could argue both ways, leaving you unsure what to think.

You might easily find yourself experiencing this dilemma. Imagine you show up at the airport one evening only to find that your flight has long since departed. The people at the check-in counter chide you, saying you should've known the scheduled departure time of 12 am meant the first thing in the morning, not last thing at night. Yet your confusion would be understandable. Officially there's no such time as 12 am. Midnight is both the end of one day and the start of the next. In 24-hour time notation, it's both 2400 hours and 0000 hours.

Aristotle appealed to a similar principle when he argued that time can't have a beginning or an end. Every moment is both the end of an era and the start of something new; every event is both the outcome of something and the cause of something else.

So, how could time possibly end? What would prevent the last event in history from leading to another? Indeed, how would you even define the end of time when the very concept of *end* presupposes time? "It is not logically possible for time to have an end," says University of Oxford philosopher Richard Swinburne. But if time cannot end, then the universe must be infinitely long-lived, and all the riddles posed by the notion of infinity come rushing in. Philosophers have thought it absurd that infinity could be anything but a mathematical idealization.

The triumph of the big bang theory and the discovery of black holes seemed to settle this question. The universe is shot through with singularities and could suffer a distressing variety of temporal cataclysms. But ask what singularities actually are and the answer is not so clear. "The physics of singularities is up for grabs," says Lawrence Sklar, a philosopher of physics at the University of Michigan.

The very theory that begat these monsters suggests they can't really exist. At the big bang singularity, for example, relativity theory says that the precursors of every single galaxy we see were squashed into a single mathematical point—not just a tiny pinprick but a true point of zero size. Likewise, in a black hole, every single particle of some hapless astronaut gets compacted into an infinitesimal point. In both cases, calculating the density means dividing by zero volume, yielding infinity. Other types of singularities don't involve infinite density but an infinite something else.

Although modern physicists don't feel quite the same aversion to infinity that Aristotle and Kant did, they still take it as a sign they've pushed a theory too far. For example, consider the standard theory of ray optics. It explains eyeglass prescriptions and funhouse mirrors. It also predicts that a lens focuses light from a distant source to a single mathematical point, producing a spot of infinite intensity. Now, in reality, light gets focused not to a point but to a bull's-eye pattern. Its intensity may be high, but it's always finite. Ray optics errs because light isn't really a ray but a wave.

In a similar vein, nearly all physicists presume that cosmic singularities actually have a finite, if very high, density. Relativity theory errs because it fails to capture some important aspect of gravity or matter that comes into play near singularities and keeps the density under control. To figure out what goes on will take a more encompassing theory, a quantum theory of gravity. Physicists are still working on such a theory, but they figure that it will incorporate the central insight of quantum mechanics; that matter, like light, has wavelike properties. These properties should smear the putative singularity into a small wad, rather than a zero point, and thereby banish the divide-by-zero error. If so, time may not, in fact, end.

Physicists can argue it both ways. Some think time does end. The trouble with this option is that the known laws of physics operate within time and describe how things move and evolve. Time's end points would have to be governed not just by a new law of physics but by a new type of law of physics, one that avoids temporal concepts such as motion and change in favor of timeless ones such as geometric elegance.

In one proposal, Brett McInnes of the National University of Singapore drew on ideas from string theory. He suggested the primordial wad of a universe had the shape of a torus. Because of mathematical theorems concerning the donut-like torus shape, it had to be perfectly uniform and smooth.

At a black hole singularity, however, the universe could have any shape whatsoever, and the same mathematical reasoning need not apply; the universe would in general be extremely raggedy. Such a geometric law of physics differs from the usual dynamical laws in a crucial sense: It is not symmetrical in time. The end wouldn't just be the beginning played backward.

Other quantum gravity researchers think that time stretches on forever, with no beginning or end. In their view, the big bang was simply a dramatic transition in the eternal life of the universe. Perhaps the pre-bangian universe started to undergo a big crunch and turned around when the density got too high—a big bounce. Artifacts of this prehistory may even have made it through to the present day. By similar reasoning, the singular wad at the heart of a black hole would boil and burble like a miniaturized star. If you fell into a black hole, you'd die a painful death, but at least your time line wouldn't end. Your particles would plop into the wad and leave a distinct imprint on it, one that future generations might see in the feeble glow of light the black hole gives off.

By supposing that time marches on, proponents of this approach avoid the need to speculate about a new type of law of physics. Yet they, too, run into trouble. For instance, the universe gets steadily more disordered with time. Now, if it's been around forever, why isn't it in total disarray by now? As for a black hole, how would the light bearing your imprint possibly escape the hole's gravitational clutches? The bottom line is that physicists struggle with antinomy as much as philosophers have. John Archibald Wheeler, a pioneer of quantum gravity, wrote, "Einstein's equation says, 'This is the end' and physics says, 'There is no end.'"

Faced with this dilemma, some people throw up their hands and conclude that science can never resolve whether time ends. For them, the boundaries of time are also the boundaries of reason and empirical observation. Yet others think the puzzle just requires some fresh thinking. "It is not outside the scope of physics," says physicist Gary Horowitz of UC Santa Barbara. "Quantum gravity should be able to provide a definite answer."

Just as life emerges out of lifeless molecules that organize themselves, time might emerge from some timeless stuff that brings itself to order. A temporal world is a highly structured one. Time tells us when events occur, for how long and in what order. Perhaps this structure was not imposed from the outside but arose from within. What can be made can be unmade. When the structure crumbles, time ends. By this thinking, time's demise is no more paradoxical than the disintegration of any other complex system. One by one, time loses its features and passes through the twilight from existence to nonexistence.

The first to go might be its unidirectionality—its arrow pointing from past to future. Physicists have recognized since the mid-19th century that the arrow is a property not of time per se but of matter. Time is inherently bidirectional; the arrow we perceive is simply the natural degeneration of matter from order to chaos. If this trend keeps up, the universe will approach a state of equilibrium, or "heat death," in which it can't possibly get any messier. Individual particles will continue to reshuffle themselves, but the universe as a whole will cease to change. Any surviving clocks will jiggle in both directions and the future will become indistinguishable from the past.

A few physicists have speculated that the arrow might reverse so that the universe could set about tidying itself up, but for mortal creatures whose very existence depends on a forward arrow of time, such a reversal would mark an end to time as surely as heat death would.

The arrow isn't the only feature that time might lose as it suffers death by attrition. Another could be the concept of duration. Time as we know it comes in amounts: seconds, days, years. If it didn't, we'd be able to tell that events occurred in chronological order but not how long they lasted. That scenario is what University of Oxford physicist Roger Penrose presents in his book *Cycles of Time: An Extraordinary New View of the Universe*. Throughout his career, Penrose seems to have had it in for time. He and University of Cambridge physicist Stephen Hawking showed in the 1960s that singularities don't just arise in special settings but really should be everywhere. He's also argued that matter falling into a black hole has no afterlife and that time has no place in a truly fundamental theory of physics.

In *Cycles of Time*, Penrose begins with a basic observation about the very early universe. It was like a box of LEGO bricks that had just been dumped out on the floor and not yet assembled—a mishmash of quarks, electrons and other elementary particles. From them, structures such as atoms, molecules, stars and galaxies had to piece themselves together step by step. The first step was the creation of protons and neutrons, which consist of three quarks apiece and are about a femtometer across. They came together about 10 microseconds after the big bang (or the big bounce, or whatever it was).

Before then, there were no structures at all—nothing was made up of pieces that were bound together. So, there was nothing that could act as a clock. The oscillations of a clock rely on a well-defined reference such as the length of a pendulum, or the distance between two mirrors or the size of atomic orbitals. No such reference yet existed.

Clumps of particles might have come together temporarily, but they couldn't tell time, because they had no fixed size. Individual quarks and electrons couldn't serve as a reference, because they have no size, either. No matter how closely particle physicists zoom in on one, all they see is a point.

The only size-like attribute these particles have is their so-called Compton wavelength, which sets the scale of quantum effects and is inversely proportional to mass. And they lacked even this rudimentary scale prior to a time of about 10 picoseconds after the big bang, when the process that endowed them with mass had not yet occurred.

"There's no sort of clock," Penrose said. "Things don't know how to keep track of time." Without anything capable of marking out regular time intervals, either an attosecond or a femtosecond could pass, and it made no difference to particles in the primordial soup.

Penrose proposed that this situation describes not only the distant past but also the distant future. Long after all the stars wink out, the universe will be a grim stew of black holes and loose particles. Then even the black holes will decay away and leave only the particles. Most of those particles will be massless ones such as photons and again clocks will become impossible to build.

You might suppose that duration would continue to make sense in the abstract, even if nothing could measure it. But researchers question whether a quantity that can't be measured even in principle really exists. To them, the inability to build a clock is a sign that time itself has been stripped of one of its defining features. "If time is what is measured on a clock and there are no clocks, then there is no time," according to philosopher of physics Henrik Zinkernagel of the University of Granada in Spain.

Despite its elegance, Penrose's scenario does have its weak points. Not all the particles in the far future will be massless; at least some electrons will survive, and you should be able to build a clock out of them. But if Penrose is on to something, it has a remarkable implication. Although the densely packed early universe and ever emptying far future seem like polar opposites, they're equally bereft of clocks and other measures of scale. "The big bang is very similar to the remote future," Penrose says.

He boldly surmises that they're actually the same stage of a grand cosmic cycle. When time ends, it'll loop back around to a new big bang. Penrose, who has spent his career arguing that singularities mark the end of time, may have found a way to keep it going. Even if duration becomes meaningless, and the femtoseconds and attoseconds blur into one another, time isn't dead quite yet. It still dictates that events unfold in a sequence of cause and effect that's the same for all observers. In this respect, time is different from space.

Two events that are adjacent within time—like when I type on my keyboard and letters appear on my screen—are inextricably linked. But two objects that are adjacent within space—my keyboard and a Post-it note—might have nothing to do with each other. Spatial relations simply don't have the same inevitability that temporal ones do.

Under certain conditions, though, time could lose even this basic ordering function and become just another dimension of space. This idea goes back to the 1980s, when Stephen Hawking and James Hartle sought to explain the big bang as the moment when time and space became differentiated. Even if you can't define duration or causal relations, you can still label events by the time they occurred and lay them out on a time line.

Several groups have made progress on how time might be stripped of this last remaining feature by studying what happens to it at black hole singularities using one of the most powerful ideas of string theory, known as the holographic principle.

A hologram is a special type of image that evokes a sense of depth. Though flat, the hologram is patterned to make it look as though a solid object is floating in front of you in 3-D space. The holographic principle holds that our entire universe is like a holographic projection. A complex system of interacting quantum particles can evoke a sense of depth—that is to say, a spatial dimension that doesn't exist in the original system.

But the converse is not true. Not every image is a hologram; it must be patterned in just the right way. If you scratch a hologram, you spoil the illusion. Likewise, not every particle system gives rise to a universe like ours; the system must be patterned just so. If the system initially lacks the necessary regularities and then develops them, the spatial dimension pops into existence. If the system reverts to disorder, the dimension disappears where it came from.

Imagine, then, the collapse of a star to a black hole. The star looks 3-D to us but corresponds to a pattern in some 2-D particle system. As its gravity intensifies, the corresponding planar system jiggles with increasing fervor. When a singularity forms, order breaks down completely. The process is analogous to the melting of an ice cube. The water molecules go from a regular crystalline arrangement to the disordered jumble of a liquid. So, the third dimension literally melts away. As it goes, so does time.

If you fall into a black hole, the time on your watch depends on your distance from the center of the hole, which is defined within the melting spatial dimension. As that dimension disintegrates, your watch starts to spin uncontrollably, and it becomes impossible to say that events occur at specific times or that objects reside in specific places.

What that means in practice is that space and time no longer give structure to the world. If you try to measure objects' positions, you'll find they appear to reside in more than one place. Spatial separation means nothing to them; they jump from one place to another without crossing the intervening distance. In fact, that's how the imprint of a hapless astronaut who passes the black hole's point of no return, its event horizon, can get back out.

"If space and time do not exist near a singularity, the event horizon is no longer well defined," Gary Horowitz says. In other words, string theory doesn't just smear out the putative singularity, replacing the errant point with something more palatable while leaving the rest of the universe much the same. Instead, it reveals a broader breakdown of the concepts of space and time, the effects of which persist far beyond the singularity itself.

The theory still requires a primal notion of time in the particle system and scientists are still trying to develop a notion of dynamics that doesn't presuppose time at all. Until then, time clings stubbornly to life. It's so deeply engrained in physics that scientists have yet to imagine its final and total disappearance.

Science comprehends the incomprehensible by breaking it down, by showing that a daunting journey is nothing more than a succession of small steps. So it is with the end of time. And in thinking about time, we come to a better appreciation of our own place in the universe as mortal creatures.

The features that time will progressively lose are prerequisites of our existence. We need time to be unidirectional for us to evolve and develop. We need a notion of duration and scale to be able to form complex structures. We need causal ordering for processes to be able to unfold. And we need spatial separation so that our bodies can create a little pocket of order in the world. As these qualities melt away, so does our ability to survive. The end of time may be something we can imagine, but no one will ever experience it directly.

As our distant descendants approach time's end, they'll need to struggle for survival in an increasingly hostile universe, and their exertions will only hasten the inevitable. After all, we aren't just passive victims of time's demise; we're perpetrators of it. As we live, we convert energy to waste heat, and we contribute to the entropy degeneration of the universe.

If time must die, it maybe be ultimately because we live.

IMAGE CREDITS

iv: Bernie_photo/E+/Getty Images ; 2: claudyo2001/iStock/Getty Images; 4: Brankospejs/iStock/Getty Images Plus; 6: OmaPhoto/iStock/Getty Images; 7: OmaPhoto/iStock/Getty Images; 8: Bet_Noire/iStock/Getty Images Plus; 22: NiPlot/iStock/Getty Images Plus; 25: gremlin/E+/Getty Images; 27: NASA; 29: maradek/Creatas Video+/Getty Images; 30: IncrediVFX/iStock/Getty Images Plus; 31: Philipp Tur/iStock/Getty Images Plus; 33: Anna Bliokh/iStock/Getty Images Plus; 49: BlackJack3D/iStock/Getty Images Plus; 51: gremlin/E+/Getty Images; 51: Austrian National Library/Wikimedia Commons/Public Domain; 54: Bryan Christie/Scientific American; 55: carloscastilla/iStock/Getty Images Plus; 56: Suppachok Nuthep/iStock/Getty Images Plus; 57: Tim Paulawitz/iStock/Getty Images Plus; 58: Philipp Tur/iStock/Getty Images Plus; 70: allanswart/iStock/Getty Images Plus; 73: Photos.com/Getty Images; 74: KevinKlimaPhoto/iStock/Getty Images Plus; 75: MarinaRazumovskaya/iStock/Getty Images Plus; 76: Melissa Thomas/Scientific American; 78: FrenchToast/iStock/Getty Images Plus; 79: gremlin/E+/Getty Images; 81: dem10/E+/Getty Images; 82: andreusK/iStock/Getty Images Plus; 98: MATJAZ SLANIC/E+/Getty Images; 100: NASA's Goddard Space Flight Center/Jeremy Schnittman; 102: estt/Getty Images; 104: Philip Howe/Scientific American; 104: dualstock/iStock/Getty Images Plus; 109: Lucy Reading-Ikkanda/Scientific American; 120: Det Kongelige Bibliotek/The Royal Library (Copenhagen)/Internet Archive; 122: Bauglir/Wikimedia Commons/CC BY-SA 4.0; 123: Rudolphous/Wikimedia Commons/CC BY-SA 4.0; 124: Gts-tg/Wikimedia Commons/CC BY-SA 4.0; 124: Maahmaah/Wikimedia Commons/CC BY-SA 3.0; 125: Rachael Towne/flickr/CC BY 2.0; 125: Mike Peel/Wikimedia Commons/CC BY-SA 4.0; 126: SilverV/iStock/Getty Images Plus; 126: Isabelle Grosjean/Wikimedia Commons/CC BY-SA 3.0; 127: Timwether/Wikimedia Commons/CC BY-SA 3.0; 127: The Museum of Science and Art, Volumes 5-6/Google Books; 128: Landscapes, nature, macro/iStock/Getty Images; 129: Deutsche Fotothek/Wikimedia Commons/Public Domain; 129: Chris Burks/Wikimedia Commons/Public Domain; 130: University of California Libraries/Internet Archive; 130: Timwether/Wikimedia Commons/CC BY-SA 3.0; 131: Mimadeo/iStock/Getty Images Plus; 132: Metadata Deluxe/flickr/CC BY 2.0; 133: Mike Peel/Wikimedia Commons/CC BY-SA 4.0; 134: MBLWHOI Library/Internet Archive; 135: Museumsfoto/Wikimedia Commons/CC BY 3.0 DE; 136: National Institute of Standards and Technology; 137: National Institute of Standards and Technology; 152: Dnn87/Wikimedia Commons/CC BY-SA 3.0; 154: Binarysequence/Wikimedia Commons/CC BY-SA 3.0; 157: Bryan Christie Design/Scientific American; 159: CNES/ESA; 162: NIST; 175: KPegg/iStock/Getty Images; 177: Sam Edwards/OJO Images/Getty Images; 178: Sergb95/Wikimedia Commons/CC BY-SA 4.0; 180: Terese Winslow/Scientific American; 181: Darwin Brandis/iStock/Getty Images Plus; 182: image2roman/iStock/Getty Images Plus; 183: Vera_Petrunina/iStock/Getty Images; 184: YassineMrabet/Wikimedia Commons/CC BY-SA 3.0; 185: BenAkiba/E+/Getty Images; 186: Ilya2k /Creatas Video+/Getty Images; 202: solidcolours/iStock/Getty Images Plus; 206: EllenaZ/iStock/Getty Images; 208: ksushsh/iStock/Getty Image Plus; 221: NASA/ESA; 223: Richard Sword/Scientific American; 226: gremlin/E+/Getty Images; 242: andreusK/iStock/Getty Images Plus; 244: Philipp Tur/iStock/Getty Images Plus; 246: Prostock-Studio/iStock/Getty Images Plus; 249: Nadia Strasser/Scientific American; 251: NADIA STRASSER; SOURCE: A AND B ADAPTED FROM FOTINI MARKOPOULOU (http://arxiv.org/abs/gr-qc/9704013/); C ADAPTED FROM CARLO ROVELLI (http://arxiv.org/abs/gr-qc/9806121/)/Scientific American; 269: vchal/iStock/Getty Images; 270: JPL-Caltech/NASA; 274: janiecbros/iStock/Getty Images Plus

NOTES

NOTES

NOTES